Michael Duffy is a publisher and political columnist. He edited the *Independent Monthly* from 1993 to 1996.

Man of Honour

MAN OF HONOUR

JOHN MACARTHUR – DUELLIST, REBEL, FOUNDING FATHER

MICHAEL DUFFY

MACMILLAN
Pan Macmillan Australia

First published 2003 by Macmillan Australia Pty Limited
St Martins Tower, 31 Market St, Sydney

National Library of Australia
cataloguing-in-publication data:

Michael Duffy
Man of Honour

ISBN 0 7329 1178 8.

Duffy, Michael. Man of honour: John Macarthur:
duellist, rebel, founding father.

Bibliography.
Includes index.

1. Macarthur, John, 1766-1834. 2. Sheep ranchers – New South Wales – Biography. 3. New South Wales – History - 1788-1851. I. Title.

994.402

Typeset in 12.5 Bembo by Cooper Graphics
Printed in Australia by Griffin Press

To my parents, Kevin and Margaret,
for giving me an interest in books of both kinds

You can have no idea of the class of persons here who consider themselves gentlemen.

William Bligh, Sydney, to the
Honourable Charles Greville,
London, 1807

my honour, much dearer to me than existence.

Lieutenant James Marshall
to a Sydney court, 1801

CONTENTS

LIST OF ILLUSTRATIONS

LIST OF MAPS

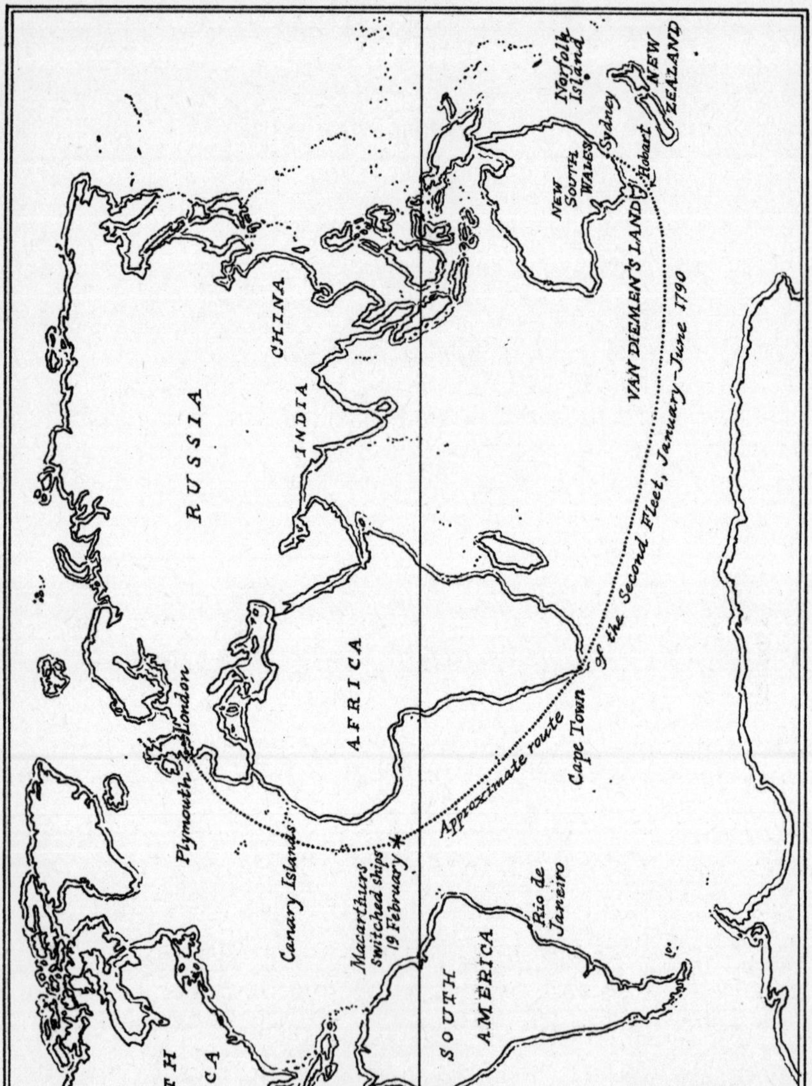
THE WORLD 1786~1811
N
EQUATOR
NORTH AMERICA
SOUTH AMERICA
AFRICA
RUSSIA
INDIA
CHINA
Plymouth
London
Canary Islands
Macarthurs Switched ships 19 February
Approximate route of the Second Fleet, January–June 1790
Cape Town
Rio de Janeiro
Tahiti
NEW SOUTH WALES
Sydney
VAN DIEMEN'S LAND
Hobart
Norfolk Island
NEW ZEALAND

Introduction

The credit for the creation of Australia is widely shared, but the most important person among its founders was John Macarthur. His wife Elizabeth and he were the first couple to decide to come to Australia on their own initiative. He played a leading part in establishing the wool export industry, which for many years dominated Australia's economy and helped shape the national character. He owned ships and traded around the world, using his sons and nephews, and his patrons in England, to advance his interests. When he was in exile in Britain on two occasions, Elizabeth ran the farms, thereby becoming Australia's first businesswoman of stature. Macarthur was a formidable amateur lawyer, and involved himself continually in what passed for politics in the colony. He helped depose Governor William Bligh in 1808, an event that came to be known as the Rum Rebellion, and later became a member of the first legislative council in Australia. A talented controversialist of amazing vehemence, Macarthur fought three duels, and almost fought several more. Gradually he went mad, spending much of the last two years of his life in confinement. By any standard, his life was extraordinary.

It's true that other men had made the decision to settle New South Wales, and on a day-to-day basis it was usually the governor who had the most power in the colony. But from the mid-1790s Macarthur was always one of the next most powerful men, and acquired an immense cumulative importance and influence. British politicians and public servants mattered, of course, but they came and went. All the governors left Sydney after a while, sometimes with Macarthur's help, but he remained.

And so did his story. Thanks in part to the survival of extensive and accessible family papers, we know more about the Macarthurs than we do about other early settlers, and this alone makes them important for us. Their original house, Elizabeth Farm at Parramatta, is today open to the public. It is now in the middle of Sydney, and part of it is the oldest European building in Australia. There is continuity, too, from the fact that their descendants still live on the land near Camden granted to Macarthur in 1805, in the mansion he built there. So there are tangible connections between him and us that don't exist for his contemporaries.

This book is about the first two-thirds of Macarthur's life, up to the Bligh rebellion and its aftermath. In wanting to know more about Macarthur and this period I have been driven by a feeling that the history of early Sydney had more influence on the future of Australia than is often acknowledged. If it is accepted that Australia has been distinguished by the extent of its egalitarianism and secularism, and its combination of vibrant democracy and strong, even paternalistic, government, it is not hard to see the roots of such characteristics in early

Sydney, provided we can look beyond the convict experience, so popular in recent decades that it has tended to obscure other views.

In early New South Wales the need for expertise was so pressing that convicts soon came to be judged on their abilities and their behaviour, not their past crimes. Ex-convicts – emancipists – were treated for most purposes as equal to men who'd arrived here free. The first governor, Arthur Phillip, gave the same rations to everyone, from convicts to himself, a disturbing and exciting innovation. His successor, John Hunter, told London not to send free men to fill certain positions in the colony, because convicts could do it just as well. Ex-convicts were given land and convict labour, and some became very wealthy. Governor Lachlan Macquarie invited them to dine at Government House. Most of these actions were radical – and much criticised by some, including Macarthur – and helped develop a society more egalitarian than Britain's.

Because New South Wales was set up as a large prison-cum-government department, government played an important role in the society that developed, and this has continued until today. We are on the whole not a nation that prides itself on self-reliance, but one that values its citizens' right to persuade government to intervene in and improve most areas of life. Due to the pervasive role of government from the start, religion was unable to establish itself here with the success it had back in Britain, where its role as a provider of welfare and social services, such as hospitals, gave it a public importance it never attained in this country, which became one of the most secular on earth. Again, all this began in 1788.

The origins of other aspects of Australia can be found in those early days, and I'll consider some of them in the chapters that follow. Their influence persisted in Sydney and spread to the later colonies, directly and, more importantly, via Britain. The experience of early New South Wales – at times much debated in Britain – had a strong influence on the attitudes and decisions of London regarding subsequent colonisation issues.

How did John Macarthur relate to all this? He opposed egalitarianism, but by striving for wealth, independence and political power for himself, he contributed to the reduction of the autocratic powers of the early governors. However, Macarthur was to learn, as the English aristocracy had learned, that the dilution of authority, once begun, could not be halted at some arbitrary point, with freedom for the wealthy only. It would develop into self-government and, in the case of Australia many decades later, one of the world's earliest and more successful democracies. Macarthur helped in this, even though his motivation was to help himself. Adam Smith suggested the role of the *unseen hand* in economics; there is an unseen hand in the growth of democracy too.

One of the first requirements in understanding historical figures is to grasp something of the most important ways in which their world differed from ours. In Macarthur's case, one of the key differences is the way his age dealt with conflict between powerful men. Today there is a range of highly structured mechanisms for enabling males to get along with each other and to help them resolve their arguments. Team sports are essential. Playing them is important for obvious reasons, and so too is being a spectator, an activity that satisfies a great deal of the male need for conflict in a generally non-violent manner. Schools try to teach boys about suppressing anger and delaying gratification. Later in life, business activities soak up violent tendencies in ways that are generally socially beneficial, as do the legal system and parliament. These last two institutions are usually portrayed in terms of their contribution to the spread of human rights and freedom; their effectiveness in stopping men from killing each other and laying society waste as a side effect is often overlooked.

While all these institutions existed in Macarthur's period, they were primitive, restricted, and far less effective. This suggests, given the many successes of Georgian Britain (the first four Georges ruled from 1714 to 1830), that its powerful men must have used some other method to rub along with each other. And they did. This was a way of thinking and behaving known as the code of honour, which determined who was a gentleman and how gentlemen should behave, particularly when in conflict with each other. It is difficult to describe the code precisely or comprehensively. It was largely unwritten: in the tradition of British empiricism, the code was not codified. But we know it existed and was widely accepted and followed, not least because it frequently led to duels, events whose reality is not in doubt. According to various contemporary estimates, there were maybe several dozen duels a year in Britain. While many readers will regard the institution of the duel with a combination of derision and abhorrence, it actually – in the absence of alternative methods – served a valuable social purpose by restricting the effects of masculine aggression and violence.

As members of the most materialistic and secular generations in history, most of us can understand the financial aspects of Macarthur's life without much effort. We are used to following money trails. What requires more effort from us is to appreciate the other values of that life, which, although often related to material interests, had an independent existence and could in certain circumstances predominate.

The code of honour stood on a line between medieval chivalry and the ideals seen in *Tom Brown's Schooldays*. It has not completely disappeared. In the twentieth century its manifestations included the notion of *parole*, when an officer gave his word not to escape, and the cold-shoulder given to business men who survived the sinking of the *Titanic* by taking space in the lifeboats that 'ought' to have been given to women and children. Basically, to restrict it to its British manifestation, the code was the ideology of a type of man called the gentleman. To be a gentleman meant many things. He could read and write and knew the classics, treated women well, dressed appropriately and displayed good manners, did not lie and was brave. He might be a military or naval officer or own land, and ideally did not work in trade, at least not in retail. As this brief list suggests, what it meant to be a gentleman and follow the code of honour was fluid and not always definitive – there were plenty of farmers, for instance, who could read and write, treated women well, and did not lie, but who were not gentlemen. And while the code of honour was an ideal of good behaviour, it was also a mechanism for resolving disputes in duels using violence, with an outcome that was often arbitrary. Indeed, the code was sometimes used as a cover for bad and dishonourable behaviour: the mafia connotation of this book's title is deliberate. Like many gentlemen, John Macarthur lived by the code but he also abused it on occasions for his own benefit.

However, these qualifications should not blind us to the reality that the code *worked* for Macarthur and his contemporaries. The duel was a way of localising violence and preventing a dispute spreading to fights between the family and friends of those involved. Historically, the code came after the vendetta and before the triumph of the rule of law. In Georgian Britain and New South Wales, two major ways of settling disputes between powerful men – the code and the law – existed in parallel and often clashed. Most accounts of the growth of the law pay little attention to what it replaced, which leaves us ill equipped to understand certain behaviour in the period of transition. The code largely disappeared around the middle of the nineteenth century, and little has been written about it.

John Macarthur cannot be fully understood without acknowledging the important role that a sense of honour played in his life and that of the other powerful men in the colony. It contributed significantly to the overthrow of Governor William Bligh. Although there is no contemporary statement that the events of January 1808 were caused specifically by the code, they were preceded and followed by an enormous amount of talk of honour and what it meant to be a gentleman. In this book I take these words seriously, and see where they lead.

The so-called Rum Rebellion was the climactic episode in early Australian history. It involved most of the powerful men in the colony, seriously affected the lives of some of them, and caused the British authorities to think seriously about New South Wales for the first time since they had decided to settle it, back in 1786. It led to the withdrawal of the New South Wales Corps, whose officers comprised over half the colony's upper class, and the appointment as governor of Lachlan Macquarie, who changed Sydney in many ways, not least by the encouragement he gave to emancipists. The popular view of the Rum Rebellion is that Macarthur and other leading rebels saw Bligh as a threat to their economic interests, most importantly to their profits from trading in alcohol. This view is partly due to the prescriptive nature of the name Rum Rebellion, which – unlike the names of other upheavals, such as the French Revolution and the Eureka Stockade – not only labels an event but interprets it. In fact this is not accurate: rum was hardly an issue, and historians demolished this view of the Bligh rebellion long ago, even though it persists in the popular imagination, along with wild notions of the rapaciousness of Macarthur and the other rebels. Most contemporaries did not blame it on rum (although Bligh tried), and the name Rum Rebellion was only devised many decades later, in 1855, by teetotaller and Bligh-enthusiast William Howitt.

While the rebels certainly resented Bligh's interference with their wealth and power, this is not a complete explanation of the rebellion. John Macarthur and Major George Johnston, the military leader, were intelligent, well-established and wealthy settlers, familiar with the ways of the world. Each of them had a family and lived in a state of domestic happiness. They must have been aware that the chances of their material interests being furthered by an armed rebellion against a governor appointed by the King of England were uncertain, to say the least. In the event, Macarthur had to leave New South Wales for over eight years, Johnston was court-martialled and thrown out of the army, and the officers who had supported him were dragged away from New South Wales, where they had become accustomed to a level of comfort and fine weather not usually enjoyed by the British army. And this was a good result – those involved would have known when they acted that they risked punishment and even death. A legal opinion obtained by the government after the rebellion recommended the prosecution of several of those involved, although eventually only Johnston was charged. Lachlan Macquarie, on his way to New South Wales in 1810 after the rebellion, believed Johnston would be executed for his part in it.

What this suggests is that there must have been some other reason for the rebellion, one sufficiently powerful to push those involved to act in a manner which, in the absence of that reason, would have been uncharacteristically reckless. This book proposes the code of honour as the missing element. The code contributed in a small way to the initial dispute between Bligh and Macarthur, and then helped widen it, drawing others into the argument and directing their behaviour. In a pattern seen in many of Macarthur's earlier disputes, the code encouraged him to make his grievance particularly personal. Bligh being the governor, it was not proper for Macarthur and him to fight a normal duel, so they became enmeshed in a convoluted, complex shadow fight, drawing in political ideas and other people's problems, and involving New South Wales' court system, so primitive and ineffective that it resolved nothing and made things worse. The affair involved theatre, and its culmination, when Bligh was deposed at gunpoint, resembled a sort of displaced duel. Like the duel proper it involved rituals, posturing and guns. As was the case in most duels, no one was hurt. It has seemed somewhat absurd to future generations, just as the duel proper seems absurd. But that is only because we have not fully understood what was going on.

While much of early Sydney did persist in later Australia, the code of honour did not. Indeed, it died more quickly here than in Britain, because of New South Wales' lack of tradition and growing sense of egalitarianism. When Macarthur returned from exile in 1817, New South Wales' men of honour were giving way to men of paper, people more recognisably like ourselves, lawyers, journalists, public servants, businessmen and, eventually, politicians. Macarthur adapted to these changed circumstances, with some difficulty but effectively. The rest of his life was fascinating but very different from what had gone before. So this book concludes with the Bligh rebellion, that great turning point in early Australian history.

*

My main sources are Macarthur's letters and speeches, and reports of his conversation. In describing and understanding the life in which these were written and occurred, I am indebted to the work of a number of historians. These are named later, but here I give particular thanks to the three men who have written about Macarthur at the greatest length, M.H. Ellis, H.V. Evatt, and Alan Atkinson, and to George Parsons' observations on the New South Wales Corps and honour.

John Macarthur signed his name McArthur for the period covered by this book, but I have used Macarthur, the spelling he later adopted. In all quoted material, which is set in italics, I have modernised the spelling and use of capital letters, in order to remove the unhelpful air of quaintness these tend to convey to the modern reader. Obvious misspellings and questionable grammar have been allowed to stand, without the use of [sic], as they tell us something of their writers.

I

Voyage

The best piece of luck an ambitious person could have in eighteenth-century Britain was to be born into a well-connected family. Patronage played an enormous and pervasive role in appointments to positions throughout society. This was something John Macarthur began without. His parents, Alexander and Catherine, were from Scotland and had possibly lived in North America for a while before settling in Plymouth Dock, around 1760. This was a new area built to service the docks of Britain's great harbour, from which James Cook's three voyages to the South Seas departed. Alexander was a linen draper, and provided sailors' clothing and other goods to the navy. The family lived in a large house at 77 Fore Street that had once housed the commissioner of the dockyard, and Alexander seems to have been moderately successful.

John was born on 13 August 1766, the younger of two surviving sons, and was probably educated at a grammar school near Exeter run by a Cambridge scholar. He boarded, either at the school or nearby. The curriculum emphasised the classics, particularly the study of Roman history and poetry. The authors read by schoolboys of the time included Caesar and Livy, who espoused heroic virtues demonstrated in wars of invasion. Cicero's work taught British boys about the links between politics and what went on in the courts. Then there were Virgil and Horace, who wrote on the attractions, indeed the moral benefits, of farming. Slavery and empire are accepted, if understated, features of such works; urban life is not among their chief preoccupations. The contradictions between the republican ideals expressed by some of these authors and the imperial circumstances in which they were writing were dealt with skilfully, and would have provided a useful template for some of the moral complexities facing the builders of the British Empire.

Macarthur was born at a time of enormous opportunity. Britain had made more progress towards democracy than almost any other nation. Its creative spirit and economy were booming thanks to the scientific and industrial revolutions and the expansion of the empire. The material and social prospects for an ambitious man of Macarthur's background were better than they had ever been. Capitalism had come of age: in 1776, when Macarthur was ten, Adam Smith published *An Inquiry into the Nature and Causes of the Wealth of Nations*, which was widely acclaimed and read by educated men, in Britain and later in New South Wales.

John's mother died in 1777. He left school at fourteen and in 1782 became an ensign (the most junior officer rank) in an independent company of foot soldiers based at Plymouth Dock. The American War of Independence was under way and there were fears the French might take advantage of this to invade Britain, even though the two nations were not at war. John's company had been formed specifically to guard against any such invasion. It was one of the most humble outfits in the army, an appropriate entry point for someone of John's modest background. The commission cost £400 and it's doubtful his father could have raised this. It is possible John borrowed the money from his brother, James, who had married the daughter of a successful local maltster and property owner.

The army of Georgian Britain was very different from ours. It was one of the few means of social and sometimes financial advancement for men of middling birth. Becoming an officer automatically made such a man a gentleman and therefore, at least in theory, an equal in some ways to every man above him on the social scale. Young officers looked forward to action and foreign service, offering as they did not just excitement but the chance of promotion should superior officers succumb to bullets or disease. That the young officers themselves ran similar risks was not overlooked. But the danger was acceptable because there were so few opportunities for advancement in their society. It was far less of a meritocracy than ours, so thrusting and intelligent men had to take big gambles in the search for some way of gaining an edge over their fellows.

Military service also offered the chances of making contacts with useful people, of commercial activity and even of receiving land grants. The experience of British officers in North America over the previous century showed how one might resign one's commission abroad, receive land and build a better life. Some officers in India were making enormous fortunes from trade. Altogether, there were enough opportunities associated with the military for a commission to be a good start in life for many young men. So John joined up, but five months later the American war ended and his company was disbanded. He turned seventeen and was now effectively unemployed and on half-pay, which meant he remained an ensign and received a modest income from the government but had no duties.

According to Macarthur family tradition, he spent the mid-1780s living on a farm near Holsworthy in Devon. The west country was an important area for wool, one of Britain's major industries. Devon was full of sheep and just to the north were the towns where wool was processed before being sent to the mills. We know little about what Macarthur did during this period. According to his son, he *took a lively interest in the various rural occupations going on, hunted, being a bold and accomplished horseman, and spent much time in the perusal of such books as he could obtain – especially works on English and Roman history and the general principles of the law. At this time he contemplated leaving the army, and going to the bar.* The English history he read would have emphasised noble actions, and have been composed in many cases to provide moral precepts for the behaviour of its readers. There would have been an emphasis on character and narrative.

Britain had been exporting convicts to North America since Tudor times. It had only three responses to crime: death, transportation, or whipping. (Gaols were largely holding pens.) Many people believed that criminals were a class of person, or even what we would call a personality type, genetically predisposed to steal, and passed on their characteristics to their children. So transportation, by removing them from the nation, was a form of eugenics. With America no longer available since the War of Independence, the numbers of criminals in county and city gaols began to increase, and overflowed into hulks on the Thames and elsewhere. In 1777 a man named John Howard had written a book exposing the sordid conditions in the nation's gaols. He was an evangelical Christian, one of a growing group keen to reinvigorate the Anglican church. They saw criminals not as a problem to be exported, but as souls to be redeemed. To do this more effectively, Howard proposed that Britain build prisons to house its criminals at home. For most people the idea seemed eccentric: one of the main aims of transportation was to exile the criminal personality; there was little belief in the possibility of rehabilitation. Also, transportation was cheap, because the convicts sent to North America had been sold to employers there for a fixed period. Prisons, on the other hand, would be expensive to build and run. So in 1786 the government decided to send its convicts to New South Wales.

The pro-penitentiary enthusiasts were not happy, and were to seize any excuse to criticise the colony for decades to come. Among them was William Wilberforce, another evangelical. A man of gentle birth, he had taken to religion in a big way in 1785, and for the rest of his life would advocate a range of good causes, with considerable success. Like most humanitarians, he had a horror of duelling, and would soon launch the campaign to shut down the British slave trade. Wilberforce persuaded his friend, Prime Minister William Pitt the Younger, to send an evangelical, the Reverend Richard Johnson, as chaplain to the First Fleet.

Evangelicalism could be described as a sort of mini-reformation of the Church of England. Back in the early eighteenth century, John Wesley had introduced a more severe, more Bible-focused form of religion to the lower classes. There was little enthusiasm for this among the gentry in the established church, so Methodism was formed, with a reputation for preaching hellfire and damnation. Now Wilberforce and other reformers began to introduce a milder version of these beliefs into Anglicanism. At first this was treated with scorn by most middle- and upper-class Britons. Apart from anything else, Methodism was somewhat egalitarian, leading one Duchess of Buckingham to reflect: *It is monstrous to be told that you have a heart as sinful as the common wretches that crawl upon the earth.* But the evangelicals, who were fervent and missionary in their zeal, persisted, and their numbers and influence grew, as we shall see.

During the lost five years of Macarthur's life, there's some reason to think he worked as a part-time teacher at Kilkhampton Grammar School near Bude in Cornwall, and that through this he met Elizabeth Veale, who was a friend of the family of one of the pupils at the school. John was now twenty-two years old, with small eyes, high cheekbones and proud, somewhat fleshy lips. His face was marked by the effects of smallpox. He had a powerful voice and was intelligent and pushy. Elizabeth had been born on 14 August 1766, and so was one day younger than him. She lived in Bridgerule, a small village on the River Tamar, which flows along the border between Devon and Cornwall down to Plymouth Harbour. Her father, a farmer, had died at the age of thirty-two, when Elizabeth was six. Her mother later remarried, and after the death of her second husband, married again. Elizabeth spent a lot of her childhood with her maternal grandfather and with the family of the local vicar, the Reverend John Kingdon, an Oxford graduate and respected scholar.

Elizabeth's childhood, while fairly tumultuous to most modern eyes, was a reasonable one by the standards of her time. At least she survived it, which was an achievement: her sister had died of illness at the age of two, and the Kingdons' five-year-old daughter Griselda was drowned in a sudden surge of the river through the middle of the village in 1786.

Elizabeth was also fortunate, as a girl, to have received something of an education from John Kingdon, which would naturally have involved the Bible. The Old Testament is rich in its depictions of the pastoral life, alive with patriarchal figures and herds of sheep. Job, for example, had thousands of sheep and engaged in agriculture, growing vines and making dairy products. His family and their neighbours fought the neighbouring tribes; there were cities in their world, but they did not live in them. There were also slaves, and in this world, unlike Elizabeth's own, they were often freed.

Despite her intelligence and her learning, Elizabeth had no property or money, and would be expected to leave the vicarage at some point and enter a raw and dangerous world. Much would depend on her choice of a husband, which was limited by the narrow range of suitable males available locally. And so she met John Macarthur. He was considered proud and, given his relative poverty and lack of opportunities, haughty. She saw something in him, though, later writing of *the activity which was very early discernable* in his mind. Elizabeth herself was regarded, she said, as *indolent and inactive*, although this might have referred to nothing more than a fondness for books and a dislike of the activities – good works and the like – associated with vicarage life.

She would have been attracted by how John regarded himself as a gentleman. This didn't guarantee a splendid future, but it was the first step for anyone seeking to become a member of the gentry. A man having this status might be considered for certain positions and responsibilities that would otherwise be closed to him. A gentleman had to be regarded with some respect by other gentlemen. If he was treated as a member of the lower classes – if, for example, he was assaulted or accused of lying – his status of gentleman was threatened. This was why the institution of the duel was necessary. The offended gentleman could – indeed, for the sake of his honour he must – demand satisfaction in the form of a duel.

But John's pressing need was to find work, and in April of that year his fortune changed and he was gazetted as an ensign in the 68th or Durham Regiment of Foot. He exulted: *A guinea a week and another three for every recruit I can find.* It's possible he achieved this through a friend of the Kingdons who was related to the London agents of the 68th. If so, it's an example of how contacts were so important at that time. Elizabeth might have brought no money to her relationship with John, but the transfer to the 68th, by putting him back on full-pay, was worth a great deal. The regiment was based in Gibraltar, but John remained in England, travelling around to recruit.

In May 1788 John and Elizabeth were close enough to stand as godparents to the Kingdons' new daughter, who was named Elizabeth. About a month later they were close enough to conceive Edward, their first child. In October 1788 the vicar married them in his own church. That Elizabeth was pregnant at the time of her marriage did not hold the negative connotations it came to acquire during the nineteenth century. In sexual matters, at least, the morality of the Georgians was closer to our own than to the puritanism of the long Victorian period, introduced by the evangelists, which was to take hold in the 1830s and run until the 1960s. The vicar put the description *Esq.* after Macarthur's name in the marriage register. While this term was vague, it tended to be used to refer to a member of the landed gentry, which John was not. Its use suggests that he had ambitions and had convinced his new friends he might achieve them.

John was reluctant to join his regiment in Gibraltar. Possibly he had transferred to the 68th because he wanted to leave the army and it would be easier to sell a commission in this regiment than in his original one. Or perhaps he did want to pursue a military career, for financial reasons, but felt there were insufficient opportunities at Gibraltar. Although now on full-pay, he was absent without leave for most of 1788. While extended leave was not uncommon in the army, Macarthur overdid it to the point where his commanding officer requested the War Office to force him to join his regiment. The young ensign responded by asking for permission to sell his commission, but this was refused in March 1789. Macarthur and Elizabeth set off for London in the mail coach and their first child, Edward, was born at Bath on the way. In the capital, Macarthur tried three times to have the decision reversed, which must have tried his superiors' patience. He was told he could neither sell his commission nor remain in Britain on full-pay, although he could return to half-pay if he wished.

The Bounty, South Pacific Ocean

While Macarthur was fuming in London, a naval commander named William Bligh was suffering his own problems on the other side of the world. On 28 April 1789 a mutiny occurred on HMS *Bounty* and its captain and eighteen of his men were cast off in a longboat. All but one made it to Timor some time later. The main purpose in referring to the affair here is to note that the causes of the mutiny, even though it is one of the most discussed incidents in naval history, are still uncertain. Originally it was believed that Bligh had been a brutal commander, but this is not true. There's no doubt the mutineers were keen to return to their Tahitian mistresses, but whether this alone would have caused them to risk arrest and the death penalty is unlikely. So this mutiny shares some of the mystery of subsequent events of 1808 in Sydney.

It has often been observed that Bligh used bad language in his dealings with subordinates. His defenders have been quick to point out that many naval commanders swore, and to reject this as a significant contributing cause of the mutiny. Yet Bligh did not just swear; by all accounts he had the capacity to make men feel small by the way he talked to them. His comments were received as vicious, demeaning, and personal. They seem to have undermined his listeners' sense of their own status, indeed, their very sense of themselves. It must have been the tone as well as the content, for afterwards some men would complain about him almost tearfully, and yet on the page the words he used do not seem to us to have justified such a response.

But in some cases they were clearly brutal. Three weeks after leaving Tahiti with its cargo of breadfruit, the *Bounty* had anchored at the island of Annamooka, and on 25 April Bligh's deputy, Fletcher Christian, was sent ashore in command of a watering party. He had trouble keeping the natives at a distance and some items of equipment were stolen. According to one observer, Bligh later *damned him for a cowardly rascal, asking him if he was afraid of a set of naked savages while he had arms; to which Mr Christian answered 'The arms are no use while your orders prevent them from being used'*. After a similar incident a while later, Bligh told the men they were *a parcel of lubberly rascals* and thrust a pistol at one man and *threatened to shoot him for not paying attention*. On 27 April Bligh became enraged about some stolen coconuts, and when Christian asked if he thought he'd taken any, replied, *Yes you damned hound, I do. ... God damn you, you scoundrels, you are all thieves alike ... I'll make half of you jump overboard before you get through Endeavour Straits.* Such comments went well beyond swearing.

When the mutiny was taking place and Bligh asked for its reason, Christian, the leader, said to him, *Not a word sir, or you are dead this instant.* As the men were being forced into the boat, Bligh again asked for a reason, and again was told several times, *Hold your tongue sir, or you are dead this instant.* In his recollection of the event, Bligh said he continued trying *to change the tide of affairs*, while Christian *continued to threaten me with instant death if I did not be quiet.* Perhaps the men simply didn't feel like conversation, or didn't want to or couldn't put their reasons into words. But it's also possible their responses meant exactly what they seem to mean: that they were profoundly weary, psychically exhausted, by the sound of Bligh's voice and the effect it had on them.

According to the journal of James Morrison, who stayed behind on the ship, Bligh said at the last minute, *I'll pawn my honour, I'll give my bond, Mr Christian, never to think of this if you'll desist.* To which Christian replied, *No, Captain Bligh, if you had any honour, things would not have come to this.* When another man tried to pacify Christian, he replied, *'Tis too late, I have been in hell for this fortnight past and am determined to bear it no longer. And you know Mr Cole that I have been used like a dog all the voyage.*

The evidence does not suggest Christian was treated like a dog by the standards of the day. But he felt that he had been, and the others supported him, and some of them might have felt similarly themselves. The *Bounty*, unusually, lacked both commissioned officers – other than Bligh – and marines, so authority depended more than usually on convention. Fletcher Christian and some of those engaged in the mutiny were gentlemen by birth; William Bligh was not. In abusing them so contemptuously and thoroughly for the entire voyage he was threatening their sense of identity. (In general the ordinary seamen seem to have been less bothered by his abuse.) It must have been a cruel experience for them, yet one difficult to put into words because so unusual, and one for which there was no redress, as the slights were delivered by Bligh in his official capacity as their commander.

The social elements in the mutiny were noted by some contemporaries. Sir John Barrow, who in 1831 wrote the first history of the *Bounty* affair, noted that *Seamen will always pay a more ready and cheerful obedience to officers who are gentlemen than to those risen to a command from among themselves. It is the common observation in the service that officers who have risen before the mast are generally the greatest tyrants.* Bligh, the son of a customs official, had not been born of a particularly 'good' family.

So neither sex nor violence explains this mutiny adequately. Sex was important, but the gentlemen's ability to resist its attractions might have been weakened not just by their long sojourn at Tahiti but by the way Bligh had undermined their sense of themselves. Somehow, by the way he spoke to them, William Bligh seems to have eaten into the souls of the gentlemen on the *Bounty*. The mutiny was about status and respect, and reflects their importance in Georgian Britain and its empire.

London

In late June or early July 1789 John Macarthur had a breakthrough: he was offered the position of lieutenant in the New South Wales Corps, which was being formed to go to Sydney and replace the marines sent out with the First Fleet.

The settlement at Port Jackson had been founded by Governor Arthur Phillip in January 1788, following James Cook's charting of the coast of New South Wales in 1770. Cook's voyage had received extensive publicity and there had been considerable public discussion in the intervening years regarding the suitability of New South Wales as a place to which convicts could be transported. The Macarthurs would have been aware of this, reading in the newspapers of the growing clamour for government action. In March 1786, for instance, the *Edinburgh Magazine* published a petition from the Sheriffs of London to the King, pointing to *the rapid and alarming increase of crimes and depredations in this city and its neighbourhood.* Since transportation to America had ceased, it was claimed that crime rates had shot up because of the loss of exile as a deterrent. The Sheriffs begged his majesty for *a speedy and due execution of the law, both as to capital punishment and transportation; without which … the mischiefs complained of must daily and rapidly increase.* The petition was directed at the government of William Pitt the Younger, and, Britain being a democracy of (the wealthier) property owners, that government eventually felt it necessary to act.

The Macarthurs might have been aware that the First Fleet was by the standards of the time a triumph of organisation and seamanship. However, the first year of the settlement had not been easy – for one thing, the government had underestimated the amount and variety of stores needed – and the reports that were published from the colony were often negative. On 23 April 1789, the *Morning Post and Daily Advertiser* published a letter from an anonymous gentleman in Sydney, almost certainly a marines officer, noting that the country was virtually unuseable: *it will never answer the intentions of government, and I make no doubt but we shall be recalled, which I sincerely wish. … we had a great deal of thunder and lightning, which has done some damage, and a few days ago we had a slight shock of an earthquake. In short, I believe this country to be the outcast of God's works.* But opinions varied. In July a correspondent of the *Edinburgh Magazine* claimed that *The soil, in general, is black, and fit for any grain sown, or roots planted, in it.*

The British were curious about Australia, and some half a dozen of those on the First Fleet had publishing contracts before they left England. The first book about the new colony was published in London in the same month the Macarthurs had to make their decision. In his *A Narrative of the Expedition to Botany Bay*, marines Lieutenant Watkin Tench predicted that

Speculators who may feel inclined to try their fortunes here will do well to weigh what I have said. If golden dreams of commerce and wealth flatter their imaginations, disappointment will follow.

Governor Phillip's reports to the government were generally upbeat, although he did request more supplies. The trees were thicker, and the sandstone harder, than had been expected. Illness used up medicines sooner than had been hoped. In his planning back in England, Phillip had had to guess about matters such as their likely success in agriculture, based on reports by James Cook and Sir Joseph Banks that proved to be overly optimistic. As 1788 progressed, their guesses (or the amounts of supplies approved by government) proved wrong, and Phillip asked London for assistance. As soon as the request was received, the government acted. In April 1789 it was decided to send the *Guardian*, a fast Royal Navy frigate, to the colony. It was packed with supplies and carried a number of convicts with experience of farming and gardening. The ship sailed on 8 September, with 1,003 tons of materials. It was hugely expensive and was a virtual ark, containing enormous quantities of food, clothing, equipment, animals and plants. There were 100 axes and 4,000 hooks; the plants included grapes, artichokes, horseradish, tarragon, almonds, mulberries, nectarines and peaches; there were stallions, bulls, cows and sheep. All up it cost £70,000, and would have suggested to John and Elizabeth that the government was determined to make a success of New South Wales, an important consideration given some abandoned colonising efforts in Britain's past.

Although the early years were difficult, the colony was more successful initially than many people today realise. Phillip was able to write in September 1788 that *those who have gardens have vegetables in plenty and exceeding good in kind.* Elizabeth wrote to her mother in October 1789 how she'd heard that in Sydney *that wheat which had been sown, flourished in a manner nearly incredible, and that the settlers are making rapid progress in building.* We don't know whether this was based on reports in the press, private information acquired by the Macarthurs, or whether Elizabeth was just fibbing to reassure her mother. But it seems likely the Macarthurs, like other people considering joining the New South Corps, had some source of reasonably positive information about the prospects of the colony.

They lived in a society where people were used to thinking about emigration, where many people had relatives and friends abroad, and had even, as is possible with John's parents, been there themselves. The example of America was always before them. We don't know if John had read Adam Smith's *An Inquiry into the Nature and Causes of the Wealth of Nations*, but Smith was an enthusiast for globalisation, which is what the empire was in part. He had written that *The discovery of America, and that of a passage to the East Indies by the Cape of Good Hope, are the two greatest and most important events recorded in the history of mankind. Their consequences have already been very great: but, in the short period of between two and three centuries which has elapsed since these discoveries were made, it is impossible that the whole extent of their consequences can have been seen.* More practical, in terms of the decision John and Elizabeth now faced, was Smith's confident claim that *The colony of a civilised nation which takes possession, either of a waste country, or of one so thinly inhabited, that the natives easily give place to the new settlers, advances more rapidly to wealth and greatness than any other human society.*

Beliefs such as these would have influenced the Macarthurs' thinking about New South Wales. If they went, Elizabeth would be only the second woman of her class to go to Australia. (The first was probably the wife of the Reverend

Richard Johnson, although almost nothing is known of her.) The journey out alone posed many dangers, not least shipwreck or death from disease. On the other hand, their options were limited in Britain. It's true that parts of the country were booming, and had been for the past fifty years. In the Midlands, men such as Matthew Boulton and Josiah Wedgwood were using new scientific discoveries to push technology and manufacturing forward as never before. The era of steam was dawning, as was that of more sophisticated financial systems. The effects of all this were transforming life in some parts of England, but not the world of farms and docks where John and Elizabeth were born. In a sense they were being left behind, as the modern world was created elsewhere, in London and to the north, by people more educated, inspired and connected than themselves. If they wanted to improve their chances they would have to move, to a place where they would be higher up the competitive scale.

Service in a new colony on the other side of the world might be tedious and difficult. But this could make it less appealing to more experienced or better connected officers, thereby creating opportunities for people such as the Macarthurs. If they, spurred by poverty, were prepared to stick there, John might well rise quickly as older officers grew bored and came home. And there was, perhaps brightest of all, the hope of land. Such hopes had to be balanced against the odium that might come to a gentleman and his wife who willingly chose to go and live among criminals and expose their children to the threat of moral corruption.

In the end, John and Elizabeth decided to take a chance, to risk their lives and that of their child and leave England. John transferred into the New South Wales Corps with the rank of lieutenant in June 1789 on a salary of £79 a year, plus a share in the recruiting bonus for every man who joined.

When Elizabeth informed her mother of the decision, she acknowledged that at first she had been against the idea, *blinded by common and vulgar prejudices*. But had she stayed in Britain, we would never have heard of her.

It was probably a very hard decision to make. That John and Elizabeth did make it suggests they were tough – they'd certainly seen enough deaths between them already – ambitious, even desperate. Contacts probably played a part. When they were boys, John's older brother, James, had struck up a friendship with Evan Nepean, who had risen to be Under-Secretary of the Home Office. Evan was the most important public servant involved in the settlement of New South Wales, and had placed his own younger brother, Nicholas, in the corps as a captain. Now John was given a place as one of Nicholas' lieutenants.

Elizabeth told her mother that John and she *had every reasonable expectation of reaping the most material advantages* by leaving England, adding that *You will be surprised that even I, who appear timid and irresolute, should be a warm advocate for the scheme*. It was a bold decision, and perhaps gave Elizabeth more confidence in herself, and helped bind their marriage together more tightly. *I have not now, nor I trust shall ever have one scruple or regret*, she added to her mother. She was talking herself into it, creating what in a sense was to be the pure private world of the Macarthur family amidst the moral chaos of the society they were about to enter, as they departed the green and pleasant land of Devon, with its streams and vicarages, for a life among the exiled scum of Great Britain.

At about this point in most accounts of Australia's settlement it is customary to pause and reflect on the enclosure acts that were forcing British peasants off their land and into the city slums. There they were chewed up by the factory hells of the industrial revolution or induced by alienation, unemployment and gin to commit the crimes for which they were transported. This is all true, and important, but there were other things happening in Georgian Britain that were at least as influential in shaping Australia. Britain was a very wealthy and productive nation, and its wealth was more evenly spread than in most countries. Visitors from the continent frequently commented on the prosperity of the lower orders, their healthy and plentiful diet and the cleanliness of their inns. In 1784 the French aristocrat Rochefoucauld noted how the English ate more meat than any other nation. All ranks of society drank heavily but worked hard. There was a widespread obsession with agricultural improvement. George III contributed articles to the *Agricultural Magazine* under a pseudonym, and many English aristocrats took a deep interest in farming that shocked their effete cousins on the continent. There was more social mobility in Britain than in most European countries. Lord Eldon, Lord Chancellor for almost all the period from 1801 to 1827, was the son of a coal-factor's apprentice. While the House of Lords was hereditary, the Commons was a democracy, albeit of the wealthy only. Governments did change. There was a press. Public opinion affected the decisions of government. Talented and original members of parliament did achieve prominence.

In Britain the law was more important than anywhere else, and did more to set limits on power. Anyone was entitled to a fair trial, and often got one. Government officials were bound by the law and judges were appointed for life. Trials were in public and by jury. The standard of proof favoured the accused. The law of habeas corpus ensured that people could not be detained without good reason. We take such things for granted today, and call them rights, but for the Britons of the 1790s they were still fresh enough to be valued intensely and consciously. They were part of what made Britain superior to other nations, and helped it avoid excesses such as French absolutism and French revolution.

It was Britons' mistrust of authority that explains one of the peculiarities of the time so relevant to early Australia: the absence in Britain of an organised police force. The population wanted no armed government officials who might be used by the monarch against them. The standing army was kept to an absolute minimum for this reason, and the people were prepared to endure a constant crime wave rather than agree to a police force which (as was the case elsewhere in Europe) could have done something about it. One of the reasons punishments were so severe was that the chance of being caught was so remote. When a criminal was seized, an example had to be made. The historian E.P. Thompson has observed that *the ritual of public execution was a necessary concomitant of a system of social discipline where a great deal depended upon theatre.* The theatre of the gallows.

The British believed in the sanctity of private property because it gave those who owned it independence from the monarchy. Therefore the laws defending property were upheld with an enthusiasm that can seem brutal to us, but which involved political ideas about liberty as well as material self-interest. To understand a great deal of what the Georgians thought and did we must remember that they were still looking over their shoulders at the Civil Wars of the previous century, and the abuses by the monarchy that had produced them.

Another important aspect of the time was the craze for trade. Britain was in the process of creating an empire. Huge fortunes were made during the eighteenth century in America, India and the West Indies, from slaves, coffee, tea, sugar, cotton goods, steam engines and other items. Trade was greed, and also creativity and innovation and excitement. And the trading urge permeated all levels of society, right to the top, to the younger sons of the aristocracy and into the armed and diplomatic services. Where most of the European upper classes regarded trade with disdain, many British aristocrats embraced it with enthusiasm.

These, then, were some of the things that shaped John and Elizabeth Macarthur and explain a lot about how they were to act in Australia, the sort of activities they believed worth engaging in, and the type of society they sought to build. The people who settled Australia are sometimes judged using standards based on what has happened since their time; but they, of course, were the product of what had happened before it.

1789

The First Fleet, which had left England with about 770 convicts in May 1787, consisted of nine transports and two navy vessels. The Macarthurs sailed in the so-called Second Fleet. This name actually encompasses several departures in 1789 and 1790, beginning with the *Lady Juliana*, which left in July 1789 with female convicts only. In September, as already noted, the *Guardian* followed, jam-packed with stores. The Macarthurs sailed in January 1790 on one of three convict transports, accompanied by another storeship. Their vessel was the *Neptune*, which at a supposed 809 tons was almost twice as large as the others, the *Scarborough* and the *Surprize*. (It was later suggested in a government report that the *Neptune* was in fact much smaller than the 809 tons claimed by its contractors.) The *Neptune*'s captain, Thomas Gilbert, had sailed with the First Fleet and written a book about the voyage, published in late 1789.

Altogether, some ninety-two soldiers went out with these ships, with the remaining two hundred members of the New South Wales Corps to come later. The Macarthurs embarked in London on 13 November 1789. The scene at the dock would have been chaotic: there were over 1,500 ships on the Thames just between Limehouse and the Tower, not to mention swarms of small boats. John's main duty was to command the soldiers who guarded the ship but had little to do with the convicts directly. For this there were specially employed supervisors, who would give out the food and arrange for convicts to be put in and out of irons and moved onto the deck for exercise. Floggings were carried out by the sailors. Basically, the soldiers' role was to keep watch and act only when violence was threatened.

The arrangements surrounding the *Neptune*, as was usual with convict ships, involved both the government and private enterprise. The transportation of convicts on the First Fleet had been contracted by the government to William Richards junior, a humanitarian who had done the job well, with few convicts dying en route. However, this had been achieved at considerable cost: £54,000 for 770 convicts. Major Robert Ross, Governor Phillip's deputy, observed from Sydney that *it will be cheaper to feed the convicts on turtle and venison at the London tavern than be at the expense of sending them here*. Determined to save money this time, the government, although it used William Richards for the *Lady Juliana*, accepted the lowest quote when it put the rest of the Second Fleet out to tender: just £22,370 for 1,250 convicts. The winning bid came from the slave-trading company of Camden, Calvert and King. They were hard men, and not only to people in irons. In 1771 Thomas King, then just a captain working for Anthony Calvert on a slave-buying voyage, had violently kicked a sailor named John Warren who was rude to him. Warren began to spit and piss blood, but was forced to do physical labour for several days in the tropical heat, and collapsed and died. King was later charged with murder, bailed by

Calvert, and found not guilty on a technicality at the Old Bailey. None of this stopped him later going into business as a partner with Calvert, or prevented the government from dealing with their company. The arrangement for the Second Fleet was for so much a head to be paid by the government for each convict taken on board – whether they lived to see Sydney or not. So the company would profit if convicts died and no longer had to be fed.

After the soldiers were loaded in late October, the first of the convicts were taken on from Newgate Gaol and from the hulks. The *Neptune* was the only one of these three ships to take female as well as male convicts. Stores were brought on board, including bread, flour, salt pork and beef. There was 400 gallons of vinegar (disinfectant and mouthwash), soap, pigs and probably some cattle, sheep and poultry, with food for all these animals.

The *Neptune* also carried the second free settler to New South Wales, one D'Arcy Wentworth. An impoverished member of the Irish ascendancy, he was an ex-soldier and surgeon who had been charged four times with highway robbery. He had been acquitted four times too, but his influential relatives had decided that a spot of exile would be good for D'Arcy and for the family's reputation. There is no record of any contact between this handsome celebrity and the Macarthurs on the voyage out, possibly because he took up with a convict woman, which would have put him beyond their social pale.

Due to the insolence of its captain, John and Elizabeth suffered considerable privations on board the *Neptune*. This situation reflected the enormous power held by people in authority in those days, no matter how irrationally, even tyrannically, they might exercise it. The head of any enterprise, whether a business or a ship, often had almost absolute power, with those below him – even his second-in-command – being relatively powerless, unless they were well connected. It was an age without unions, or much in the way of government regulations, investigative journalism and professional associations. In this case it was Thomas Gilbert, as the ship's captain, who held final command, but the situation was complicated by the presence on board of the 26-year-old Captain Nicholas Nepean, the senior army officer. The Nepeans were the sons of a Plymouth shipwright, therefore men of roughly the Macarthurs' class in origin. Under-Secretary Evan Nepean had elevated himself through sheer talent in the public service and was described as *slow yet sure – calm and of few words – sober and laborious*. His brother Nicholas seems to have been less sure: sometimes he spoke up for his men when in conflict with Gilbert, and sometimes he did not. It was a difficult situation, and conflict was frequent.

Nepean had been given a large room to share with the Macarthurs but, on being offered more congenial accommodation for himself elsewhere on the ship, agreed for this room to be divided by a flimsy partition. In one half were lodged a number of loud and smelly female convicts; in the other the Macarthur family and their young female servant. It is a mystery of the whole business why this situation, which appears to have been so unreasonable for the Macarthurs, was allowed to occur and to persist. No other officers appear to have suffered as they did.

In his account of events, Surgeon John Harris, an Irishman in his early thirties and Macarthur's colleague in the corps, recalled how, while sailing from London to Plymouth, Macarthur complained to Captain Nepean *that it was impossible to live in his cabin owing to the stench of the buckets belonging to the convict women of a morning*. Nepean mentioned this to Gilbert, who flew into a rage (he had already dismissed previous complaints) and said *he was sure Macarthur meant to be troublesome, and that he would write to the War Office and have him turned out of the ship*. Macarthur, who was nearby, informed Gilbert that he was *an insolent fellow*. Gilbert then pushed Macarthur on the chest, and the lieutenant had to be restrained by Nepean from taking matters further while they were under sail. Tight discipline needed to be maintained over a shipload of convicts distressed by the prospect of exile. Even when gentlemen disagreed, it was understood this must not undermine their shared interest in keeping the lower orders in their place.

On account of an error in navigation, the ship arrived at Land's End under the mistaken impression it was Plymouth, and it took three days to beat back to that port. Once they arrived, Macarthur publicly attacked Gilbert, who was perhaps feeling a little vulnerable, taxing him *with his ungentleman-like conduct towards him and his wife*, and telling him that *he believed him to be a great scoundrel*. To which Gilbert replied that he *had settled many a greater man than him*. There was now only one way to resolve the dispute.

And so, in the mid-afternoon of 27 November 1789, Macarthur and Harris came down the ladder of the *Neptune* and were rowed across the grey waters of Plymouth Harbour. Once ashore, they walked over to the Fountain Tavern, where Macarthur stood off to one side while Harris talked to a man named Nelson. These two were the seconds, whose job it was to arrange the duel and see it was conducted according to the code of honour. It was 4.00pm. Gilbert stood nearby, also waiting. Harris and Nelson spoke briefly and then, when it was clear no reconciliation was possible between the two principals, the four men, walking in two pairs apart, carrying two cases of pistols, repaired to the seclusion of the Old Gun Wharf.

The pistols were loaded by Harris and Nelson, and the principals were given one each. A distance of ten paces was stepped out, and Macarthur, who was twenty-three, and the older Gilbert took their positions. For a moment they stood facing each other in the gloom of the winter afternoon. They would have been wearing dark clothes to reduce their visibility, and have stood side-on to each other. If a man presented his whole chest and was hit, the wound might not be so serious, but the chance of being struck was much greater. A signal was given – probably one of the seconds dropped a white handkerchief – and the men fired. Macarthur's ball struck Gilbert's coat but not his body, while Gilbert's shot whizzed past Macarthur harmlessly. There was a pause while the seconds reloaded the guns. Macarthur and Gilbert fired again, this time missing each other completely.

Due to the inaccuracy of pistols and the emotions natural to the occasion, such an outcome was common. According to the code of honour, this was the point at which a duel should be concluded, and Macarthur now closed off the affair by declaring his antagonist's behaviour *in every respect that of a gentleman and a man of honour*. The problem of the accommodation on the *Neptune*, which had led to the argument, had not been resolved, but there was nothing else he could do. Following a duel, the men involved, assuming they survived, were expected to live at peace with each other. The institution's purpose was not justice but dispute resolution. So the pistols were packed up and the four men left the wharf. Macarthur returned to the ship and his wife, whom he had not told about the duel beforehand.

This was the first of quite a few affairs of honour in Macarthur's life, and the first of three occasions on which he would fight a duel. He fought more than most, but the fact of fighting was not unusual. Even an apparently modern man such as Adam Smith accepted the need for this way of settling differences, writing that *Affronts in company [i.e. to one's honour] are most atrocious crimes. The trifling fine of five or ten [pounds, even when the law imposes it] is by no means adequate compensation for them. Where the law denies justice we are naturally led to take it ourselves. This introduced duelling in Europe.* Smith acknowledged that the problem arose from the seriousness with which society chose to regard insults such as *a blow, or the shaking of one's fist at one ... pulling one's nose ... giving one the lie, or other reproachful words.* He observed that some societies, such as ancient Greece and Rome, did not regard these affronts so extremely, but in countries such as Britain where *the laws of honour are received*, the duel was necessary.

So different from our world and our concerns are honour and its implications that we need to leave Macarthur for a while to dig back into the history of the subject. Duelling had been introduced into Britain in the late sixteenth century to limit the violence that occurred when differences between individuals spread to family and friends, as in *Romeo and Juliet*. It was therefore a progressive development, replacing assassination and the vendetta. While there was a vast range of reasons why men might differ, then as now, these differences, once they began, tended to be expressed as affronts to honour.

Some monarchs supported duelling, believing it encouraged martial values, but others considered it an affront to the king's law. In Scotland, for instance, the Duels Act 1600 made engaging in a duel a capital offence, and constables were empowered to bring anyone intending to fight a duel before a justice. The Duels Act 1696 made the giving, sending, or accepting of a challenge in Scotland punishable by banishment. None of this seems to have curbed the practice.

In 1622 in England James I, concerned at the increase in

duelling, revived the Court of Chivalry to punish insults between gentlemen, which often grew out of disputes over status. For example, was the mayor of Exeter, a merchant, justified in claiming to be *at all times and in any place as good a man* as Sir Popham Southcote, nominally his superior (because a knight) but also, problematically, a soap-maker? Punishments were fines and public confessions – one person found guilty of claiming to be a gentleman when he wasn't had to confess publicly that he had not respected *the gentry laws ... of this kingdom.* The court was shut down by the Long Parliament in 1641. Lord Clarendon told parliament that recently this court had awarded greater damages for *contumelious and reproachful words* in two days than had been given by the rest of the courts put together. After the restoration of 1688, Charles II supported a bill to restore the court's powers to resolve *the contending about place and precedency*, again with the intention of reducing the number of duels, but the bill did not make it through parliament.

So duels continued to be fought about honour, and honour continued to be about status. In his 1732 book *An Enquiry into the Origins of Honour and the Usefulness of Christianity in War*, Bernard Mandeville wrote that the man of honour must above all else not accept any *action designedly done to undervalue him.* Mandeville believed that calling a man of honour a liar was the worst affront possible, and Joseph Addison agreed. Writing in the *Spectator* in 1794, he said: *One may tell another he whores, drinks, blasphemes, and it may pass unresented; but to say he lies, though but in jest, is an affront that nothing but blood can expiate. The reason perhaps may be, that no other vice implies a want of courage so much as the making of a lie; and therefore telling a man he lies, is touching him in the most sensible part of honour, and indirectly calling him a coward.*

The idea of honour can seem a little absurd to us, but it

proposed a set of ideals that were admirable enough, and gave men something to aspire to. It held that men of high social status were expected to be leaders and to behave better than their inferiors, which was no bad thing. The code of honour was important for a society that depended on physical bravery in its soldiers, who often engaged in hand-to-hand combat. According to the Duke of Wellington, *The British army is what it is because it is officered by gentlemen … men who would scorn to do a dishonourable thing and who have something more at stake than a reputation for military smartness.* This was one reason why monarchs frequently overlooked duelling in their armies. (It was not successfully banned from the British army until the 1840s.) Martial values, of course, were particularly needed during the Napoleonic Wars, and in a vast prison camp such as New South Wales.

Despite its widespread acceptance, duelling was illegal. However, it was rare for a protagonist to be convicted, even if he killed his opponent, provided it could be shown in court that the duel had been conducted according to the requirements of the code of honour. Juries, like the rest of society, tended to condone duelling between gentlemen. The sort of people who served on them were keenly interested in questions of status. One of the strengths of British society was that it permitted more social mobility than most continental ones. In particular, it was relatively easy for many men to become gentlemen. This was why the pistol was used rather than the sword to settle affairs of honour. In 1765 the novelist Tobias Smollet noted that this *puts all mankind upon a level, the old with the young, the weak with the strong, the unwieldy with the nimble*. It meant that a nouveau gentleman who – unlike an aristocrat – had not learned fencing from childhood might take on any other gentleman with a chance of survival. It was also less dangerous: duels with the sword usually ended only when someone was cut, whereas in most pistol fights neither man was wounded, let alone killed. This was what society wanted: the point was for a gentleman to demonstrate courage by facing his opponent's gun. Bloodless outcomes were encouraged (although by no means guaranteed) by the practice of making duelling pistols with bores that were smooth, not rifled, and therefore less accurate.

One might assume that duelling was most common among the French, or the hot-blooded Mediterranean nations, but in fact it was the British who had the reputation for fighting over their honour. We shall see how one of the Spaniards who visited Sydney in the 1790s observed with disfavour the British enthusiasm for violent argument. And the British in their turn disapproved of other nations' reluctance to fight. The following comment was made in 1818 by a Briton returned from Spain: *There is a coarseness of manners among the higher ranks very visible in these parties, and language sometimes passed which in other countries would lead to serious consequence. To call a man a liar, or even to take him by the nose, would not produce a duel, nor perhaps be thought of the next day; the point of honour is not observed; there is in consequence none of that delicate sensibility which characterises the gentlemen in England.* Along the same lines, Mandeville believed that it was worthwhile *to see half a dozen men sacrificed in a twelvemonth to obtain so valuable a blessing as the politeness of manners, the pleasures of conversation and the happiness of company, in general.*

The world of the Macarthurs was one in which many of the props of civilisation we take for granted were simply absent. It was a harsh place, where early death was common, a world without much trust between strangers, where you were very much on your own, apart from friends and family. The relatively polite and compassionate way in which most people in the West today deal with each other was largely absent in a society where, especially for those in the middle ranks trying to better themselves, the hot breath of failure was always on their necks. There were few government services or safety nets. Genuinely public life occupied a smaller part of existence, while private life – especially family and other contacts – was far more important. In such a world, anyone wanting to protect themselves and their loved ones had to do things we don't do. Duelling was one of them. It was, if you like, a privatised form of justice, and was used in situations where antipathy, if left to grow, might fester and spread to other people in the group. Even if one did go to law (although most duels were fought over insults to which the law was irrelevant) it might take months or even years. The duel was a way of bringing matters to a head quickly and resolving them one way or another. This was far more important to society than any abstract notion of justice. Of course duelling was grossly unfair, and favoured the brave and the bully.

Elizabeth heard of John's duel on his return, but only *from some distant hints*. Her recorded reaction reveals relief, but no sense of the horror with which most modern readers consider the idea of duelling. Indeed, it's likely she would have been proud of her husband. Not only had he demonstrated courage, he had also achieved an important social milestone. Only gentlemen could duel, therefore by duelling he had proved himself a gentleman.

The next day Captain Nepean took the keys to the convicts' quarters from one of the ship's officers, and Gilbert reacted in fury, complaining to his superiors in London and trying to have the locks on the doors changed. This reflected a general state of unresolved confusion about who had ultimate responsibility for the convicts on board. On 1 December Nepean, no doubt thinking of the prospect of another six months in Gilbert's company, wrote to his brother describing the captain as *the greatest scoundrel that was ever created. His insolence to me has been so extraordinary that words can hardly convey you an idea of his infamy*. He begged his brother for help, but Evan did not interfere.

One night when Nepean and Macarthur were both on shore in Plymouth, Surgeon Harris, the senior member of the corps on board, was called by the guards to deal with a hysterical female convict. They were just about to put the hand-irons on her when Gilbert *hollow'd down the hatchway several vulgar phrases – who was that below making such a racket, damning his eyes, &c. I told him it was me, on which he dam'd me who I was; I told him my name was Mr Harris, on which he said Mr Harris might be dam'd – who gave me authority to command in his ship*. There is the suggestion in the surgeon's account that Gilbert had been drinking most of the afternoon.

As Captain William Bligh had demonstrated just seven months before on the *Bounty*, sea captains were very sensitive about being respected. So, Harris recorded, Gilbert came roaring *down the ladder, took hold of me by the shirt-breast, and dam'd me and Captain Nepean several times, as I told him I acted by his orders, and again dam'd Nepean – who gave him orders! then ran into the women's room, call'd out to them they were used in a most cruel and barbarous manner, on which I went and reason'd with him on the improper manner in which he acted in endeavouring to raise dissentions in the ship*.

Then things warmed up, for Gilbert grabbed a guard by the neck and the soldier *desir'd him not to shove him on his post, or he would run him through*. At which Gilbert ripped open his own shirt, revealing his chest, and *dam'd him why he did not run him through according to his orders*. Gilbert then disappeared and returned with his officers, who had also been drinking. They were persuaded by Harris to retire to the deck, where further lively discussion ensued, with Gilbert telling the surgeon he was *a fellow of dam'd low subterfuge.*

The ship's officers then took out muskets from the arms-chest, loaded them and retired to their cabin, while Gilbert set out in a boat to complain to a superior officer on shore. While he was away, Nepean and Macarthur turned up together with Elizabeth, who had spent two nights with her friends at Bridgerule, which is fairly close to Plymouth. She had wanted to stay longer, but came back because Gilbert had said he would be sailing that night, even though it later became apparent he had no intention of doing so.

According to Elizabeth, *We had no sooner arrived on board than a complete scene of uproar and confusion presented itself. Captain Gilbert had insulted a sentinel at his post, and struck him.* Nepean sent Harris to his brother's office with a brief message: *There is mutiny in the ship.* Gilbert appealed for help to the Plymouth Port Admiral, who appointed a mediator. Then, perhaps realising that he had offended the brother of a powerful man, Gilbert wrote to Under-Secretary Evan Nepean implying the dispute was the fault of officers other than his brother, *an officer who I beg leave to say I have, as well as the world, a very high opinion.* But it was too late for sycophancy. Evan Nepean told an acquaintance that Gilbert was *an exceedingly troublesome fellow, and I am rather surprised that my brother has not kicked him.*

Before the small convoy departed, Gilbert was replaced by Donald Trail, another of the toughs employed by Camden, Calvert and King. He was promoted from the *Surprize*, where there had also been disputes between the army and the sailors. Captain William Hill of the New South Wales Corps observed of that ship's officers: *it is impossible to suppose that men bred up gentlemen must be obliged to herd against their inclinations with a set of men who for the most part are so illiterate and uneducated as not to be able to write their names – so that they cannot behave with good manners from a total incapacity*. These disputes between soldiers and sailors reflect a pattern we will see often enough during the first half of John Macarthur's adult life. The lack of clearly defined areas of authority, and different service cultures, produced frequent disputes which became personal as men accused each other of being dishonourable – of not being gentlemen – which led to violence.

For the sake of peace on this occasion, Harris was transferred to the *Surprize*, perhaps as a sort of swap with Trail. It appears that Gilbert, although a vicious soul, might have been right in resenting Nepean's efforts to take control of the convicts. But being in the right did not avail him against such a well-connected opponent. It was a lesson in the benefits of influence that John Macarthur might have reflected on as he and his family tried to sleep accompanied by the stench and bad language emanating from the other side of the partition in the ship's grand cabin.

The remaining convicts had been taken on at Plymouth, making a total of 424 men and 78 women. Most of the men were accommodated in cramped spaces without portholes, chained by the wrists or legs and ankles. One account suggests the shackles were the same as those used in the slave trade. The convicts were provided with reasonable clothing, but their living conditions were appalling. Three large tubs served as toilets, and according to one crew member, *being frequently overset by accident or negligence, rendered the place very nauseous, and together with the stench arising from the breath of such a number of persons in so small a spot, the smell of their provisions and other unwholesome things, made it almost unbearable.*

The government's agent for the voyage was Lieutenant John Shapcote, who was about fifty years old. His job was to ensure the captain fulfilled the terms of the contract between the government and the ship's owners, particularly regarding the wellbeing of the convicts, but also of the soldiers. When he boarded the *Neptune* he had the convicts' possessions searched, and between seventy and a hundred knives and other iron implements with lethal potential were confiscated. Either Shapcote or Trail ordered that a considerable quantity of the convicts' belongings be thrown overboard, due, it was said, to the risk of infection from *gaol fever*. By the time the *Neptune* left Portsmouth, up to sixteen convicts were already dead. There were stories that the corpses had been put into the sea without their canvas shrouds being properly weighted, and four bodies had floated to shore.

The Macarthurs' accommodation was not changed by Trail, a 44-year-old Scotsman who had served in the British navy during the American War of Independence. There were other gentlefolk aboard – Nepean, Shapcote, the surgeon, and Trail and his wife – but, Elizabeth recorded, *we seldom benefited by their society*. The Macarthurs deeply resented Nepean's action in allowing the great cabin to be shared with convicts. They were also upset by his subsequent lack of sympathy for their situation, with Elizabeth musing that *in this instance, as in so many others, reason unassisted by power, proved unavailing*. So they kept to themselves in their room, *determined patiently to submit to the unpleasantries we could not remedy*. There was access to the deck by a dark corridor shared by the convicts (which they avoided using for health reasons) and a more exclusive passage from their gallery into the upper cabin.

The small fleet finally left British waters on 17 January 1790. In the Bay of Biscay a storm blew up and Elizabeth *began to be a coward*, at sea for the first time. *I could not be persuaded that the ship could possibly long resist the violence of the sea which ran mountains high*. As soon as this was over, Edward became very sick, as did the Macarthurs' maid, who caught the fever that was spreading among the female convicts. Elizabeth was terrified: *I had hourly every reason to expect that the infection would be communicated to us, as our apartments were so immediately connected with those of the women.* The *Neptune* was a very sick ship – the worst of any convict vessel in Australian history.

Elizabeth's journal of the voyage contains not one comment of sympathy for the convicts, who are referred to as *abandoned creatures* whose constant attendants are *filth and vermin.* There is a combination of indifference and savagery in the Macarthurs' attitude to convicts which was typical of their class and perhaps remains the most difficult of their views for university-educated and wealthier people to accept today. Her attitudes are much closer to those of less well educated people now, who are generally in support of harsher penalties for villains, far more likely to support the death penalty than their more genteel fellow citizens. The reason is the same in both cases: the closer one is to criminals and the more likely one is to be a victim of crime, the more vehement one's attitudes tend to be. And this is compounded by the feeling of helplessness that comes from being relatively powerless. Today the wealthy are far less likely to be assaulted and robbed than the poor, so they can afford to take a morally Olympian attitude to justice. This is helped by the fact their wealth makes robbery less painful.

Another difference exists between the Georgians and ourselves. We have enormous, well-funded systems of policing, justice and punishment to deal with criminals and a vast knowledge of human psychology and of how to cope with crime and its results. We have insurance affordable by most people, and a well-developed democracy to ensure our high-taxing governments respond to problems. John and Elizabeth Macarthur and their contemporaries had none of these things. Compared with us, they knew little of criminals and felt their society had almost no control over their behaviour. In other words, they were scared.

Elizabeth possibly knew little of the lives – and deaths – of the convicts on the *Neptune* as it sailed south, apart from what she picked up through the partition in her chamber. She kept to her room, or went up on deck *every fine evening*. Sometimes she would sit there with John, who no doubt shared with her the continuing frustrations of his job. The company that owned the *Neptune* starved not only the convicts but the soldiers. The men complained to Macarthur who, like a good officer, complained to his superior, Nepean, whose first response was, *Trail does every thing to oblige me, and I must give up some points to him.* As with the Macarthurs' cabin, Nepean was quite happy to give away what he himself did not need. Macarthur complained again, and was told *I will see into it.*

The incredible response to this was that the door to the Macarthurs' private passage to the deck was nailed shut on 30 January. The nominal reason given was that John had been using the passage to eavesdrop on the conversation of Trail and Nepean, a charge Elizabeth dismissed as *infamous and unfounded*. Even if this had been his habit, the punishment far outweighed the crime. What it meant was that the only exit from their cabin now lay through a dark corridor, which Nepean had ordered be made a sort of hospital for the sick convicts. This meant that Elizabeth dared not use the corridor in case she breathed in the convicts' fever or picked up their vermin – body and hair lice, bugs of different kinds – by brushing against them in her long dress. As a result, she and her sick baby were now confined in their small cabin as the ship approached the Equator and the temperature rose to levels she had never before experienced.

No language can express, no imagination conceive, wrote this young woman from the vicarage, *the misery I experienced …*

surrounded with wretches whose dreadful imprecations and shocking discourses, ever rang in my distracted ears: a sickly infant claiming constant maternal cares: my spirits failing: my health forsaking me. She was about four months pregnant. *To a person unacquainted with the innumerable insults and cruelties I was necessitated to bear with; this may appear the language of passion – resentment – or of heart desiring revenge. But it will be admitted to be the conclusions of truth, and of justice when it is known in addition to the wrongs I have already related that we were deprived of a part of our little ration and insultingly told, we should have less, if they thought proper.* John wrote to Lieutenant John Shapcote, asking for his help, but was told that he would *not quarrel with Trail for any man.*

From Elizabeth's account it appears they were persecuted for no reason, and that their persecutor was defended by Nepean. One wonders if this was the case, or if John had brought this treatment on them through some behaviour we're unaware of. Against this needs to be put the fact that throughout his military career he seems to have been a perfectly competent officer, enjoying the respect of most of the corps. Whatever the reason, now when their servant went to collect their daily ration of five quarts of water, she was watched carefully lest the seaman doling it out give her more than the exact entitlement.

So the wretched voyage continued, with John using *the only passage now left to us, often endangering a limb by tumbling over boxes and other lumber that this place was made the repository of: and frequently contracting heaps of the vermin with which it was infested.* The problems with the men's rations persisted, and on 10 February on deck a sergeant complained to Macarthur that an attempt had been made to cheat him of several of the men's meat ration. The chief mate overheard this and called the sergeant a *damned rascal.* When Macarthur protested, the chief mate abused him too. Macarthur then called on Nepean to resolve the dispute, but was told he should settle it himself. Given the still unresolved confusion over lines of command on board, this seems unreasonable, even insulting. Macarthur clearly thought it so: he asked to be moved to the *Scarborough*, and Nepean agreed. On 19 February, just six degrees north of the Equator, the two ships stopped and the Macarthurs and their maid were rowed from one to the other. *I suffered greatly from the heat*, Elizabeth recalled, *but this was an inconvenience I thought lightly of – after what I had been taught to bear.*

Perhaps due to his behaviour later in life, Macarthur has been accused of haughtiness on this voyage. But it seems the soldiers were poorly treated by Trail. There is plenty of evidence that Camden, Calvert and King and those who worked for them wanted to get every penny they could screw out of anyone who came within their power. Scurvy and fever raged through the convicts, and corpses were being tipped over the side of the *Neptune* frequently. Of a total of 502 convicts, 171 died and were thrown overboard by the time the ship reached Sydney, almost one a day on average. Some of the convicts had been brought on board in a weak state, which was worsened by the contractors' decision to sail all the way to the Cape of Good Hope without stopping. (The First Fleet had stopped at Madeira and Rio de Janeiro, bringing on board fresh food each time.) Even worse, Trail withheld food from the convicts, so that some of them literally starved to death. A sailor later recalled how *the allowance given to the convicts was not sufficient to support nature, and they were reduced to such extremities that they have eaten the poultices taken from their sores*. He claimed that often where two convicts were chained together and one died, the other would hide this fact, *concealing his companion's death, until the corpse became putrid, on purpose to receive the allowance intended for the dead man*. If the ship really was much smaller than claimed by the contractors, this would help explain the cramped conditions suffered by the convicts and the Macarthurs. Trail, as well as starving the convicts, let them up on deck very rarely, which of course compounded their various illnesses.

The sailors were also badly treated, fed a *short allowance of bad provisions* and subject to extensive violence from the officers. As for the soldiers, it is notable that Nepean was unconcerned by his own men's suffering. It was normal for British soldiers to be paid and fed appallingly (only the most desperate no-hopers joined the ranks), so it's possible the difference in the attitudes of Nepean and Macarthur in this case simply reflected Macarthur's lack of experience as an officer. He had not yet had his sense of humanity dulled by custom.

By the time the Macarthurs left the *Neptune*, they had spent almost two winter months at docks in England, which must have been incredibly cold and unpleasant for Elizabeth and her baby. They had then been just over a month at sea, the last nineteen days of it with no access to the deck for Elizabeth and Edward. But now things improved greatly. They arrived on the *Scarborough* after a suspected mutiny had been uncovered, the convicts involved being flogged severely and in some cases chained to the deck. Elizabeth's journal doesn't allude to this: her focus, as always, was on things affecting her family immediately. They shared a small cabin with Lieutenant Edward Abbott, a pleasant man about their own age. The ship's master was John Marshall, according to Elizabeth *a plain, honest man*, who had taken the ship to Sydney with the First Fleet and left his Newfoundland dog there. He made things as comfortable as possible for his new passengers, and impressed Elizabeth with the tenderness of his feelings for his wife and three children back in England. Whether he treated the 259 male convicts on board as well is less certain; seventy-three of them died. Despite this, he was considered more humane than the commanders of the two other transports, raising the possibility that the high death rate on his ship may have had a lot to do with the weakness of the convicts when they first came on board, and the decision not to stop before the Cape. On Marshall's trip with the First Fleet, not one convict had died on his ship.

Presently Elizabeth noted: *we passed a French Guinea Man [ship], bound to Martinico with slaves*. The event incurred no further comment from her, with the next lines in her journal noting *we crossed the line [Equator] on the 25th with a light wind: and on the 14th of April, after experiencing a severe gale of wind, anchored safely in False Bay*. This was about twenty miles from Cape Town's harbour, Table Bay, which was an unsafe mooring in winter. Here the Macarthurs stepped onto land for the first time in almost three months.

The country immediately around False Bay was mountainous, but with arable land nearby. There was a pretty town on the bay, its whitewashed houses decorated with green doors and windows. The Dutch East India Company was powerful, and the place was run too much like a business for the liking of the English. Captain William Hill, whom we last saw being upset about British sailors, observed acidly that *the Dutch answer minutely the character given of them by travellers; a change of clime, which is generally believed to produce a change of constitution and disposition, has not abated one tittle of their propensity to avarice. Their governors are merchants, and monopolise the whole stock of the adjacent country, and will not supply the shipping but at an advance of five or six hundred per cent.*

The British were shattered to learn that the *Guardian*, sent out before them full of stores to save the struggling colony at Sydney, was lying wrecked nearby in Table Bay. On the last leg of its voyage, on 24 December 1789, the ship had approached a gigantic iceberg – twice as high as its highest mast – to obtain ice, to be melted for water for the many plants and animals on board. But suddenly the weather worsened and the boats had trouble pulling back to the ship. A fog arose and the island of ice began to move, striking the ship, smashing the tiller, cracking some of the structure above the stern, and holing it. The crew threw overboard an enormous amount of guns, bellowing cattle, anchors and stores, and managed to get the ship off the ice. But it was sinking and about a third of the men decided to abandon ship. One boat was overturned in the heavy seas nearby. The others got away, although only one was ever seen again.

Those remaining on board had managed to drag a sail around the hull and over the hole, reducing but not stopping the leak. The ship's commander, Lieutenant Edward Riou, had his right arm broken by a falling barrel, and his ship's log shows the deterioration in his writing that followed when he had to use his left hand. They struggled back towards land, pumping and baling for eight weeks. At times the upper deck was almost under the crashing seas, and the ship, in the words of boatswain John Williams, lay *like a log on the water*. Finally, on 21 February, the *Guardian* limped into Cape Town. Those on board, according to one who saw them on their arrival, *looked like men of another world – long beards, dirt, and rags covered them*.

The next day Riou wrote a letter to the Admiralty, excusing its brevity by pointing out that the mail boat was about to depart, and also *at this instant I find it more necessary than ever to exert myself in order to prevent the ship from sinking at her anchors*. The remaining cargo was taken off, and two days before the Macarthurs reached False Bay a storm smashed the unlucky vessel on the shore. Those on the *Scarborough* would have been aware of the great loss to Sydney entailed by this disaster. Elizabeth Macarthur noted in a letter that the ship had been fitted out in Britain *at an amazing expense* and reported *it is said she is already $60,000 in debt at the Cape, and soon will be as many more.*

A small quantity of the *Guardian*'s stores were taken on board the newly arrived ships, but most could not fit and were stored on land, for a price. William Hill noted of *the unfortunate Mr Riou* that *I much fear the Dutch are taking every advantage of his situation, charging enormous sums for warehouse room and fresh provisions for the men, so that the cargo must, ere now, be insufficient to defray the cost.* Riou became something of a hero back in Britain when the news of his saving the *Guardian* arrived, but he remained stuck in Cape Town for many more months, and when he finally made it back his fame had been overtaken by the story of another post-disaster recovery by another lieutenant, William Bligh in the *Bounty*'s longboat.

At False Bay the Macarthurs enjoyed fresh meat, fruit and vegetables. Elizabeth observed the native costumes and flora, washed their linen thoroughly, and stocked up on supplies for the final leg of their journey. On 20 April she wrote to her mother that Edward *has been very sickly throughout the passage, and unless a very speedy change take place I am well convinced he will shortly cease to be an inhabitant of this world.* She had told her mother just before leaving that he had cut one of his teeth, but in fact she was mistaken, and his development continued to be slow – at the age of one year he was as large as a child of only four months. *He is very sensible, very lively*, she added, *and affords us much pleasure; but the trouble we have had with so delicate a little creature is indescribable, and I wonder my own health hath not suffered more from the attention I have been obliged to pay him.* On the positive side, though, *Mr Macarthur has enjoyed a remarkably good share of health, ever since we left England, and I trust will continue to do so.* Elizabeth was advised by the ship's doctor to build up her strength by spending some time on shore, where she boarded with a private family. She liked the Dutch houses and told her mother that *their slaves keep them remarkably clean. A man's riches are here determined by the number of his slaves. If you go to a genteel house you will see a dozen of them attending in the hall.* She signed off by asking to be remembered to her grandfather: *Tell him with my love I have not forgotten his counsel to have ever present to my mind the duty due by us to our maker.*

Then disaster struck, for *a few days before we quitted False Bay Mr Macarthur was attacked with a violent and very alarming fever; it continued to rage until every sense was lost, and every faculty but life destroyed.* Elizabeth nursed her husband as well she could, her situation made worse by the fact that Edward was also so *ill that I could scarcely expect him to survive a day*. Captain Marshall, and also one Captain Reid from another ship in the bay, helped her. When the *Scarborough* left South Africa, John was still very sick, and for the first five weeks of the voyage to Sydney one or two soldiers had to sit up with him every night. (At which point his reputation as a defender of his men's interests might have paid off. Would Nepean's men have sat up with him if he had lain dying?) We are not sure what his illness was, but it may have been a rheumatic one brought on by falling into cold water after exerting himself on a warm day at False Bay. Or he might have caught the same disease as their maid. During this final leg the 23-year-old Elizabeth, although she does not record it in her journal, gave birth prematurely to her baby in what must have been distressing circumstances. The unnamed girl died after an hour and was buried at sea.

It took sixty-one days to reach Sydney. Elsewhere in their little fleet, Lieutenant Shapcote, the government agent, was also ill. According to Captain Trail, Shapcote went to dinner one night and *eat very heartily of sallad and bread and cheese, and said he felt himself much better than he had for some time*. He then went to bed and died, and presently another body was heaved overboard in a weighted canvas shroud. As the ships continued on their route, they struck a great deal of cold and violent weather, which made exercise and cooking impossible at times. Some of the convicts were up to their waists in sea water for lengthy periods. Those still alive were often suffering from vitamin deficiency with beri-beri and the resultant paralysis, plus scurvy and protein deficiency. Anaemia and blood loss were common, and then there was disease. By the end of their voyage of six months, 267 of about one thousand convicts had died.

II

Phillip

The *Neptune* and the *Scarborough* sailed up the coast of New South Wales until they reached the long line of dirty yellow cliffs, interspersed with bright beaches, that is the face Sydney presents to the sea. On the southern head they saw a flag raised to alert the colony to their arrival. The ships zigzagged up a harbour that must have struck those on their decks as by turn impressively huge, empty of civilisation, and safe. Its islands and many bays and crannies would have appealed both to the sailors' pragmatic need for shelter and to those more sophisticated minds on board who had been educated in the eighteenth-century sensibility to landscape. They would have seen natives standing on the rocks, and their fires and canoes.

On 29 June 1790 the two ships anchored in Sydney Cove. It was bigger then, with the head of the cove up near today's intersection of Bridge and Pitt Streets. That was where the

Tank Stream, a modest creek that rose up in today's Hyde Park, entered the harbour. At low tide there were mud flats, although closer to the harbour proper the cove also contained some small white sand beaches. The 100-foot-high trees had been pushed back by the settlement, but ran along the rocky ridge to the right, from where another, small, stream entered the cove.

Looking up the cove, the Macarthurs saw a small wharf. Behind it lay Bridge Street, Sydney's main road, running from east to west across the Tank Stream. On the rise to the left of the stream sat the most imposing building in the small town, the two-storey house of Governor Arthur Phillip, with two small oaks and a number of peach trees in its garden. To its right were a few brick buildings housing the commissary, in charge of the all-important government stores, and the judge-advocate. Further to the right, once the street had crossed the stream, was the senior military officer, with the barracks up behind his residence. As the ships anchored, Hector, Captain Marshall's dog, splashed into the water and swam out to the *Scarborough*, to be joyfully reunited with his master. Nearby, the son of Lieutenant John Shapcott, who had arrived on an earlier ship, climbed on board the *Neptune,* hoping to be reunited with his father.

William Hill was impressed by *this beautiful heavenly clime. … Suffer your imagination to enter the regions of fiction, and let fancy in her liveliest moment paint an elysium, it will fall far short of this delightful weather.* The town itself was less pleasing, a scattering of primitive dwellings inhabited by men and women barefoot, ragged and starving. The action had long since moved inland to Parramatta, where the soil was better, and Sydney had been left behind. The settlement had been established for two and a half years, and was hungry due to the mediocre results of the first efforts at agriculture and the loss of the

Guardian. Its population was 591 in April; once the last of the Second Fleet ships arrived in June, this number increased briefly to 1,715. A large subsidiary colony had been established on Norfolk Island, with 418 people. In August several hundred of the new convicts were taken from Sydney to the island, where the lieutenant-governor was Major Robert Ross, the marines commander. The arrival of the Second Fleet, while it did bring badly needed stores to New South Wales, placed significant new pressures on the struggling colony by adding greatly to its population.

The man presiding over all this was Governor Arthur Phillip, a naval officer whose success so far had been mixed. The son of a steward and language teacher of German origin, he was fifty-two years old, small, and possessed a thin, prominent nose and a powerful but sharp voice. His temperament was thoughtful, persistent and reserved, and he had travelled widely and lived in different countries before coming to Sydney. He had brought a fleet of eleven ships and some thousand people safely halfway around the world and established the beginnings of a new society.

Despite this success, early Sydney suffered from some major inadequacies that might be called *ideological*. The British government had given almost no thought to the needs of the free society that would soon begin to grow alongside the prison. There was no treasury – no currency – and the legal arrangements were primitive. David Collins, a marines officer, had been appointed judge-advocate, the colony's principal legal officer. He had no legal training, and was also Phillip's secretary. He presided in the courts and administered them, and had to obey any order from the governor or one of his military superiors, even in the discharge of his legal duties: the potential for conflicts of interest and confusion was extensive. The arrangement indicates how narrowly the thinking of the government was focused. To add to the future problems, courts consisted of the judge-advocate and six military or naval officers, an arrangement that would create enormous problems once the activities of the corps brought its members, or their foes, before the court.

In the twentieth century there was seen much debate over Britain's intention in establishing the colony in New South Wales. Was it to be just a prison, or a strategic asset, or a source of raw materials? The debaters have searched diligently for evidence of motivation in the written records, but this evidence does not support any one view. My guess is that the decision was simply a solution to the urgent electoral problem of getting prisoners out of Britain as quickly as possible. The history of the first thirty years of the colony strongly supports such a view. It reveals neither clarity nor consistency nor even particular concern in the attitudes of successive government administrations to the problems of Sydney. What they appear to have wanted above all was that it continue to take more convicts, as cheaply as possible.

In the early years Phillip had to put up with a range of problems, not least the bitching of marines officers who wrote to England claiming the place was nothing but a desert and would never succeed. Quietly and constantly, with little sign of emotion, he countered the critics, noting in June 1790 that, given the problems he had faced, it was more *a matter of surprize that a regular settlement exists [at all] than that it is not in a more flourishing state*. Pressing on, he had persuaded the government to send more supplies, and established a moral society in which convicts had reason to hope for redemption in this world, through good behaviour and land grants. He was also a humane man who dealt well, in often trying circumstances, with convicts and Aborigines. Phillip's achievement was considerable, but so measured, unheroic and private (no personal papers are known to exist) that he left behind him almost no image or phrase that catches it for us.

Presumably Phillip would have greeted the Macarthurs on disembarking. There is no record of their first impressions, beyond Elizabeth's memory that *On my first landing, everything was new to me, every bird, every insect, flower, etc.; in short, all was novelty around me, and was noticed with a degree of eager curiosity and perturbation.* We know no more of what she thought of the blackbutt, forest red gum and turpentine trees, or the salt-water swamp oak in the marsh at the head of Sydney Cove.

She herself would have been regarded with much interest by those she met. They had had no contact with the outside world for thirty months, apart from La Pérouse's two ships right at the beginning and the *Sirius*' voyage to Cape Town some eighteen months earlier. None of the marines officers had brought his wife.

It took a week before the convicts were unloaded, and their appearance and condition caused widespread revulsion among those who saw them. It's possible the Macarthurs would have been unaware until then of just how bad conditions had been for the prisoners. But on disembarking it is likely John Macarthur, at least, would have witnessed the same scenes as his colleague William Hill, who later wrote the following report for William Wilberforce. His initial figures are exaggerated.

Humanity shudders to think that of nine hundred male convicts embarked in this fleet, three hundred and seventy are already dead, and four hundred and fifty are landed sick and so emaciated and helpless that very few, if any of them, can be saved by care or medicine, so that the sooner it pleases God to remove them the better it will be for the colony, which is not in a situation to bear any burthen, as I imagine the medicine-chest to be nearly exhausted, and provisions are a scarce article.

The irons used on these unhappy wretches were barbarous. The contractors had been in the Guinea [slave] trade, and had put on board the same shackles used by them in that trade, which are made with a short bolt, instead of chains that drop between the legs and fasten with a bandage about the waist, like those at the different gaols; these bolts were not more than three-quarters of a foot in length, so that they could not extend either leg from the other more than an inch or two at most; thus fettered, it was impossible for them to move but at the risk of both their legs being broken. Inactivity at sea is a sure bane, as it invites the scurvy equal to, if not more than, salt provisions. ... The slave trade is merciful compared with what I have seen in this fleet.

Captain Trail of the *Neptune* – one of those described by Hill as *low-lifed, barbarous masters* – having starved convicts to death by withholding their rations, now proceeded to set up shop in Sydney and did a thriving trade selling those rations at high prices. *My feelings*, Hill continued, *never have been so wounded as in this voyage, so much so, that I never shall recover my accustomed vivacity and spirits.*

The Reverend Richard Johnson, the colony's chaplain, went down into the hold of the *Surprize* when it arrived and *beheld a sight so shocking to the feelings of humanity, a great number of [the convicts] laying, some half, some nearly quite naked, without either bed or bedding, unable to turn or help themselves ... the smell was so offensive that I could scarcely bear it.* Extremely upset, he was then rowed over to the *Scarborough*, where Captain Marshall, who had been so kind to Elizabeth Macarthur, *disuaded* him from venturing below deck. He recorded of the convicts that *Upon their being brought up to the open air some fainted, some died upon deck, and others in the boat before they reached the shore.* Until Governor Phillip ordered the practice to stop, some of the corpses were thrown overboard and their bodies were washed up on the rocks around Sydney.

A female convict who had arrived on the *Lady Juliana* three weeks earlier wrote a letter describing the unloading of the new arrivals to the London *Morning Chronicle*: *they were almost dead, very few could stand, and they were obliged to fling them as you would goods, and hoist them out of the ships, they were so feeble; and they died ten or twelve a day when they first landed*. The dead were taken across to the harbour's north shore and buried there, although a cemetery was later established closer to hand, at the site now occupied by Sydney Town Hall.

Governor Phillip wrote to the Home Office that he would not *dwell on the scene of misery which the hospitals and sick tents exhibited when those people were landed, but it would be a want of duty not to say that it was occasioned by the contractors having crowded too many on board those ships, and from their being too much confined during the passage*. The Reverend Johnson, who visited the hospital buildings and tents, noted that many of the newly arrived convicts were too weak to turn their bodies, and *in this situation were covered over almost with their own nastiness, their heads, bodies, cloths, blankets, all full of filth and lice*. After talking to many of the patients about conditions on the ships he recorded that the greatest number *by far* were from the *Neptune*. In the seven months before the Second Fleet arrived, nineteen Europeans died in Sydney. In the following seven months, 288 died, and many more followed them. It has been estimated that one in 2.6 convicts of the Second Fleet died, compared with one in thirty-three of the First Fleet.

The Macarthurs were given a hut vacated by an officer who was being transferred to Norfolk Island. Movements such as this, between Port Jackson, Parramatta, and Norfolk Island, were to be an important part of events in their little world over the next few years. The hut was on the west side of the

Tank Stream, with the parade ground and the settlement's storehouses between them and the convicts. Further north (where Nurse's Walk is today) lay the hospital and the bakehouse, with the observatory of Lieutenant William Dawes on the point at the end of The Rocks, near where the south end of the Harbour Bridge now rests. John's duties, as on the ships, involved no direct contact with the convicts. He managed the troops who guarded the storehouses and other official buildings, such as Government House, against thieves, and stood by to enforce discipline if necessary.

The Macarthurs' maid looked after Edward, but Elizabeth had *no female friend to unbend my mind to, nor a single woman with whom I could converse with any satisfaction to myself, the clergyman's wife being a person in whose society I could reap neither profit nor pleasure*. Just why Mrs Johnson was regarded in this severe manner was not recorded, but we can guess that, as the wife of an evangelical minister, she might have been too overtly religious for Elizabeth, and opposed to both profit and pleasure.

As compensation, Elizabeth had the company of some of the marines officers, presumably the better behaved – those who had taken convict mistresses might have been less welcome in the Macarthur home. The Marine corps was an interesting choice as the first gaolers of Sydney. It was looked down on by the regular army, yet was more professional. It was possibly this, as well as the fact that Evan Nepean had served in it during the American War of Independence, that influenced the decision to send a detachment to New South Wales. Major Robert Ross, the commander of the four marine companies, was a veteran of the American war and also of Wolfe's capture of Quebec in 1759. He fought with and undermined the aloof Phillip, refusing to allow his officers to help supervise the convicts even in the early days when there were no civilian supervisors. Phillip had had to promote the more reputable convicts to the job. Ross persistently predicted doom for the colony, in 1788 claiming that *in the whole world there is not a worse country than what we have yet seen of this. All that is contiguous to us is so very barren and forbidding that it may with truth be said, here nature is reversed.* Phillip, weary of their quarrels, had sent him to run Norfolk Island in March 1790.

The marines officers who called on the Macarthurs walked through the eucalypts wearing red long-tailed coats, white trousers and black shoes. Their hats were also black, and they wore swords. The most acceptable guests were the perceptive author Captain-Lieutenant Watkin Tench and the amateur astronomer and anthropologist Lieutenant William Dawes. The two were friends, and together had conducted some of the early exploration of the country around Sydney. The Macarthurs saw Tench almost every day. If they had not read his book about the early days of the colony, they would now have been brought up to date. *Many a guard have I seen mount, in which the number of soldiers without shoes, exceeded that which had yet preserved remnants of leather*, he had written. The colony had been placed on short rations in April, and they would soon have learned that *If a lucky man, who had knocked down a dinner with his gun, or caught a fish from angling from the rocks, invited a neighbour to dine with him, the invitation always ran, 'Bring your own bread'. Even at the governor's table, this custom was constantly observed.* Every man when he sat down pulled his bread out of his pocket, and laid it by his plate. But not Elizabeth, the only woman to dine at Phillip's table in those months. The gallant governor sent her dainties from time to time, and she alone was provided with bread by her host when she and John dined at Government House.

William Dawes was *so much engaged with the stars that to mortal eyes he is not always visible*. Indeed, he had been provided with instruments and books for an observatory on the recommendation of the Astronomer Royal. Elizabeth seems to have got on with him: he was very religious, but he generally kept it to himself. She persuaded him to instruct her in the rudiments of astronomy and, when this proved too difficult for her, of botany. Gradually she acquired *some easy science to fill up the vacuum of many a solitary day*, and learned how to classify all the common plants around Sydney. They also talked about Dawes' study of the language of the Aborigines, sufficiently advanced for him to have formed the opinion that *they have the tradition of the Flood among them.*

With John Macarthur, Dawes perhaps discussed public works: as the colony's engineer, he had built gun batteries to guard the entrance to Sydney Cove, and laid out the government farm and the first street plans for Sydney and Parramatta. Building had not gone particularly well. It was not so much that the colonists had come unprepared: the First Fleet had brought tools and nails, hinges and locks and 5,448 panes of glass and 10,000 bricks. But it had been assumed they would find decent wood for building and lime for their mortar. However, the eucalypts either proved rotten when cut down, or twisted and split once they had been used to build with. Admittedly, the colonists had been in such a rush they hadn't dried the timber first. The cabbage trees were easier to work with, but the wood was soft and didn't last. There was clay but no lime, apart from what could be obtained by laboriously crushing the piles of sea shells discarded by generations of Aborigines and mixing the resulting powder with mortar. So bricks were generally held together with a mud paste. Major Ross' house, the second brick building in the colony, collapsed one day during heavy rain. But things were improving: good timber, for example, had been found on Norfolk Island. The Macarthurs were fortunate in arriving after several years of frustrating experimentation had been completed.

In a situation where merely to build a modest brick house was a challenge, town planning occurred only in fits and starts. Phillip had been given no instructions regarding the sort of town he was to build, possibly because no one in Britain had paid much thought to it. In 1788 he had planned a 200-foot-wide boulevard through the middle of the future city, which was to be called Albion, but later he instructed that buildings be erected where it was to have run. Nowhere in Phillip's planning was there allowance for commercial buildings around Sydney Cove, which reflects the general lack of interest in the future free inhabitants of the colony and what they might get up to.

While the settlers struggled to grow wheat and maize (corn), some other crops did better. In June 1790 Phillip wrote to Sir Joseph Banks that *I have about 1,000 cutting of vines now in the ground and have had a few grapes, and many as fine figs as ever I tasted in Spain or Portugal ... it is surprising to see how well the vines thrive.* The Sydney wildlife would have provided the Macarthurs with plenty of interest. In July a young midshipman and three marines were out in the harbour in a small boat when they saw an enormous whale following them. They tried to escape but it surfaced nearby and half-filled the boat with water. The men baled and rowed for their lives and, the whale having gone below, they thought themselves safe. Then – awful moment! – they felt the boat shake as the whale rose directly beneath them, lifting the vessel high and tipping it over. *The midshipman and one of the marines were sucked into the vortex which the whale had made,* Tench reported, *and disappeared at once. The two other marines swam for the nearest shore, but one only reached it.* Swimming was not a skill widely possessed in those days, even among those who went down to the sea in boats.

Eventually the last of the Second Fleet's ships departed and the colonists were left alone, waiting for the arrival of the next expected vessel, the *Gorgon*, with more members of the New South Wales Corps. This had been due to depart England soon after the *Neptune*. As the months crept by and there was no sign of a ship, the settlers grew depressed. One of the major events in the brief history of the colony had been the wreck of the *Sirius*, one of its two ships, in March of that year. And then the *Guardian* had been lost. The seas had not been particularly kind to young Sydney.

And if Sydney was starving, Norfolk Island, with little by way of supplies, might well be dead. Everyone was worried at the lack of news, and Elizabeth never passed a day without reflecting *with pain and anxiety* on the likely fate of its inhabitants. The continuing absence of further shipping pressed on their minds. *Week after week stole away, and month after month, with little diversity; each succeeding sunset produced among us wild and vague conjectures of what could be the cause of the Gorgon's delay*. In August 1790 Elizabeth and John turned twenty-four.

Elizabeth took an interest in the Aborigines, although disapproving of the way the men always walked before the women. She learned of the dreadful smallpox that had devastated the local tribe the previous winter, and of how dead and dying natives had been found in *the woods*. No one knew where the disease had come from – none of the settlers had smallpox, so the chance of it being spread by them was unlikely. It has been suggested that it had originated in northern Australia, where it was introduced by Macassan trepang fishermen, and by a remarkable coincidence reached Sydney soon after the First Fleet. One settler recalled how the disease affected the Aborigines, *their loathsome carcasses having been found, while this disorder was prevalent amongst them, lying about the beach, and on the rocks. In fact, such is the terror of this disorder amongst these untutored sons of nature, that, on its appearance, they forsake those who are infected with it, leaving them to die, without a friend at hand, or assistance to smooth the aspect of death, and fly into the thickest of the woods. Their superstition leads them to consider it as an infernal visitation; and its effects are such as to justify this idea, in some degree, for it seldom fails to desolate and depopulate whole districts, and strews the surface of the country with the unburied carcasses of its wretched and deserted victims.*

Two Aboriginal children were taken in by the Reverend Johnson and Surgeon John White to save them from the disease, which no European caught. They began to learn English, and Governor Phillip had two adult Aboriginal men kidnapped in the hope that they might be taught to communicate information regarding the interior. These two eventually made their escape but were sighted some time later at Manly Cove, where several hundred Aborigines had gathered and were feasting on a beached whale. Phillip rushed to the scene and was speared through the shoulder. Other spears were thrown and the governor had to retreat with a ten-foot pole firmly embedded in him, further lacerating the wound. Finally he made it into the boat and home, where he demonstrated his professionalism and nerve through an absence of resentment, and by getting his papers in order before allowing the potentially fatal removal of the spearhead. He survived.

Norfolk Island, July 1790

Other members of the settlements were less stoical. In July 1790 on Norfolk Island there occurred what seems to have been the first duel in Australia. Lieutenant Ralph Clark fortunately provided a description in his diary. As we have no similarly detailed record of any of Macarthur's duels, it's worth looking at this one to get some sense of how an affair of honour might proceed.

The matter began one evening after dinner. (Despite Elizabeth's fears, the inhabitants of the island were not starving.) Lieutenant William Faddy told Major Robert Ross of something one of his colleagues, Lieutenant Robert Kellow, had said. Regrettably we don't know what it was, but it must have been bad, for it disturbed Ross. He sent for Kellow, who denied saying it. This presented Ross with a difficult problem. *You have given each other the lie*, he said. *One of you must be in the wrong.*

But who? This was no minor question. Whereas today we would accept the situation, and the ambiguity involved in not knowing which of the two men had given the lie, for Georgian society this was intolerable. Telling the truth was possibly the most important part of being a gentleman. It was to resolve the sort of problem that now confronted Ross and the other gentlemen on the island that the duel had been developed. Of course it was an imperfect method, intended not to find the truth but, more pragmatically, to resolve the social conflict caused by the assertion of differing versions of it. And in the army, it was essential that officers felt they could trust one another.

The two men left and Ross asked Lieutenant Clark, who had observed the whole thing, if he'd ever heard anything like it. Clark said he hadn't, and Ross announced: *Until this matter is cleared up I will not put my feet under the same table with either of them*. It was necessary to punish the honest man in order to exclude the liar. Ross instructed Clark to inform Kellow of this decision, and Kellow, who was new to this sort of thing, asked what he could do to resolve the matter. *It is a delicate point to give one's advice on*, said Clark. Kellow replied that he was asking *as a friend*, so Clark told him *You have nothing left but to call Mr Faddy out*.

Kellow didn't seem to realise quite what should happen next, for he now told Clark bravely: *I am getting my pistols ready in expectation of hearing from Mr Faddy*. Clark gently explained that, as Faddy had given *him* the lie, it was *his* responsibility to initiate a duel. Indeed, for the sake of honour he had to act promptly. Kellow said *I will watch him as he goes out to Charlotte Field [an area of the island] and will carry my pistols with me for him to give me a meeting now or when he thinks proper*. In fact, it was proper for such arrangements to be made by a 'second'. Kellow did not know this, but he was about to learn.

He went looking for Faddy and found him just after dusk. *You sir*, he said, *have accused me before Major Ross and Mr Clark of saying words which I never did say, for which I am now come to demand satisfaction*. Faddy replied that he would consult their brother officers before giving Kellow a reply. This was an accepted procedure: due to the uncodified nature of the code of honour, a man involved in an affair sometimes had to take advice from his social peers and superiors in order to determine how to act.

The next morning before breakfast, Faddy told Kellow he would meet him, and Kellow talked to Clark about getting time off work that day. Clark offered to do Kellow's duty for him, but Kellow replied: *I wish you Clark to go with me as my friend [i.e. second]*. This put Clark in a quandary. He believed that his involvement as an observer of the affair and carrier of messages between Ross and Kellow precluded his acting as second. *I don't know how far with safety to myself I can do it*, he replied. However, it was a moot point and he would ask the opinion of Captain George Johnston. *If he thinks I can go out with you as your friend with any degree of safety … I will attend you with pleasure*. But Johnston knocked the idea on the head, saying: *Clark you cannot attend Mr Kellow with any degree of safety nor do I wish you so much as to think of it*.

Nevertheless, Clark seems to have felt a little guilty about his response to Kellow's request, for he was grateful when a further reason to reject it emerged. He returned to Kellow and asked who was to be Faddy's second, only to be told that he did not have one. Kellow then made the extraordinary suggestion that Clark act as seconds to both the principals. Clark told him this was quite impossible; indeed, the proposal was so outrageous it gave him the courage to reject Kellow's original request. Were he to be the second to both men, he explained virtuously, *Should either one of you fall I should be the person the law would lay hold of to answer to the person that was killed. By your asking me to go out in the manner you have, you have put it entirely out of my power to attend you as your friend.*

So Kellow asked who else he should approach. *If I was to ask Captain Johnston do you think he would?* The captain was asked and refused. Then Clark suggested *You are intimate with Mr Shortland why don't you ask him?* Lieutenant John Shortland was more agreeable and presently Kellow and he met with Faddy and John Altree, the doctor, to arrange the duel. Altree then officially informed Johnston of the proposed fight. This was not the action of a gentleman – Clark felt Atlee was *a damned rascal* to have told Johnston. As duels were illegal under the Articles of War, it was essential that superior officers be able to claim ignorance of them until after the event.

So now, in order to protect himself, Johnston had to order Kellow and Faddy back to camp and inform Major Ross. This forced the major himself to apply the law, even though it was he who had told the two men they must resolve the matter between themselves (which effectively meant he had told them to fight). Ross instigated an inquiry into the matter and told Clark to put Kellow and Faddy under arrest and confine them to their quarters.

Clark passed on the order to Kellow but he also had a very different conversation with Kellow's second, Shortland, explaining that a duel was still essential for the reputation of

Kellow and Faddy, and that being under arrest ought not to be considered a barrier to this desired event. *It is absolutely necessary*, he said, *for the sake of them both that the matter is settled.* Here we see a second, with the help of an older and wiser head, negotiating his principal through the shoals of the code of honour.

That afternoon Captain Johnston, Clark, and Lieutenants John Creswell and John Johnstone interviewed Kellow and Faddy and tried to determine who was lying. They failed, although Clark began to suspect it was Faddy who was telling the truth. It was decided to renew the investigation the next morning. However, later that day – 27 July 1790 – the duel occurred. Faddy had obtained a second, Stephen Donovan, the island's pilot. The men met out of town at 4.00pm and fired one shot at each other without hitting their targets. The seconds now tried to settle the affair, and the ensuing conversation turned to the original words exchanged by the two young officers that had led Faddy to make his report to Ross. Faddy repeated something Kellow had said, although not the bit that he had passed on to Ross. At this Kellow said he had not recalled saying these words until Faddy brought them back into his memory.

Shortland and Donovan pointed out that if Kellow forgot saying these harmless words, he might also have forgotten saying the words that had so concerned Faddy. Kellow *said it was very possible, but he called God to witness that he at present did not recollect them*. It was all very unsatisfactory: he seemed prepared to cast doubt on his own memory provided he did not have to admit to any wrong-doing. At this Shortland, who as second was responsible for his principal's correct behaviour in the matter, announced that he could not permit him to fire again, and everyone retired from the field. *Mr Donovan and I left them and came into town*, Shortland recalled, *they followed us*. It might have been wiser had the two principals been kept apart a little longer, for on their walk back into town Kellow retreated from the way out that had been offered to him, and which he seemed to have accepted, and asserted once again that he had never said the words attributed to him by Faddy. The two men announced that they would fight again the next morning.

When Shortland told Clark what had passed later that night, Clark advised him to have nothing more to do with Kellow, whose vacillating behaviour suggested he was lying. *Take my word, you will gain no credit in this matter*, he said. He believed Kellow had *now entered so far into [the lie] that he is ashamed to acknowledge it*. Shortland agreed, and said he would tell Kellow he could no longer act as his second. *I wish to God Faddy had shot him through the heart*, Clark mused, *for he richly deserves it*.

The next morning after breakfast Johnston, Creswell, Johnstone and Clark interviewed Shortland and Donovan about the duel, and then sent for Kellow and Faddy. This time Kellow, while persisting in his claim not to remember saying the words, *owned he had been too precipitate in asserting before Major Ross that what Mr Faddy had advanced was false*. The principals and seconds were sent out of the room and the four interrogators discussed the matter further, concluding that Kellow was a liar. Moreover, he could have destroyed both the character and the life (*by obliging him to have recourse to a pistol*) of a fellow officer. They found his behaviour in the affair to be *really infamous and derogatory to the character of an officer and a gentleman* and asked Ross not to *put us under the disagreeable necessity of doing duty any longer with a man who we think a disgrace to the corps he serves in and who we are determined never to associate with*. Ross accepted this recommendation and ordered that Kellow be suspended from all further duties and sent away on the first ship leaving for Sydney or England.

Before Kellow left Norfolk Island he tried to talk to Clark

again. This approach was refused, Clark informing him that *my brother officers would think it very odd of me to see me speaking and in company with you after what has passed.*

This affair demonstrates several important aspects of an affair of honour, particularly the seriousness with which lying was regarded and the way a gentleman's peers might be involved in determining how the code of honour should apply in certain circumstances. Kellow seems finally to have placed himself outside the code by acting as he did. This is one of the best accounts of a duel we have. Like most descriptions, it dwells on the behaviour of the principals and says nothing of their emotions.

Sydney, October 1790

At last, on 20 October, the flag was run up the pole on South Head again to indicate the approach of a ship. Elizabeth began to prepare herself to meet other women of her rank for the first time in almost a year. But as the vessel came into sight and moored in Sydney Cove, its identity became apparent: it was only the *Supply*, the colony's brig, which Phillip had sent to Batavia (modern Jakarta) for stores. Elizabeth's disappointment was modified when its commander, Lieutenant Henry Ball, visited the Macarthurs and gave her *many little comforts procured at Batavia, which were truly acceptable*. He also told them of a second-hand story that a ship sounding like the *Gorgon had been spoken to somewhere about the Equator* and was bound for Sydney. This intelligence, Elizabeth noted in a letter, *was brought to the Cape of Good Hope, and from thence to Batavia.* Earlier in the same letter she noted how another message had been sent *by way of China*. In these casual but recurring details we catch some sense of lives lived in exile, with much emotion riding on – and much thought given to predicting – the progress of letters and small ships creeping around the globe. It was a time in which every letter sent carried a part of its writer, and every letter received provided more nourishment than we can perhaps imagine today.

In early December Governor Phillip's gamekeeper, a convict named John McIntyre, was speared by Aborigines while hunting near Botany Bay. He was brought home and the spearhead, seven and a half inches long and barbed with wood and small pieces of sharpened stone, was extracted. Later he died. An enraged Phillip ordered Tench to take a party of soldiers to the area where the spearing had occurred and decapitate ten male Aborigines, the heads to be brought back to Sydney. Phillip pointed out that seventeen Europeans had been wounded or killed by Aborigines to date, but he had never ordered revenge killings as in those cases the assailants had acted *from having received injury, or from misapprehension.* He excused his own spearing on the latter ground. This time, though, he believed there was no such excuse.

This presented Tench with a dilemma. He had reason to believe that, while McIntyre might not have provoked the Aborigines on the occasion of his spearing, he had almost certainly killed or wounded some of them in the past. It is not recorded if he put this to Phillip, but he did suggest a more modest retaliation, involving the capture of just six men. Phillip agreed, saying, *If six cannot be taken, let this number be shot. Should you, however, find it practicable to take so many, I will hang two and send the rest to Norfolk Island for a certain period, which will cause their countrymen to believe that we have dispatched them secretly*. The posse, containing William Dawes and other marines, and Captain Hill of the New South Wales Corps, set off, carrying various supplies including *hatchets and bags to cut off and contain the heads of the slain.* Two surgeons also came along, possibly to perform the decapitations.

After the party had left, a friendly Aborigine named Colbee turned up at Government House and happened to mention to Phillip that he knew what was going on. He also said he was on his way to Botany Bay, presumably to warn the tribe. Why Phillip didn't just lock him up is a mystery. As it is, Tench's account of what happened next is one of the strangest stories from the early years of Sydney. First Phillip tried to dissuade him by offering him presents not to go – a blanket, a hatchet, even a jacket. When this didn't work, it was determined to *eat him down*, by giving him his favourite food, *of which it was hoped he would feed so voraciously as to render him incapable of executing his intention. A large dish of fish was accordingly set before him. But after devouring a large horseman [snapper] and at least five pounds of beef and bread, even until the sight of food became disgusting to him, he set out on his journey with such lightness and gaiety as plainly showed him to be a stranger to the horrors of indigestion.*

The expedition failed to find any Aborigines who would allow themselves to be either shot or captured, so Tench was sent out again, this time with Lieutenant Edward Abbott and Ensign John Prentice of the New South Wales Corps. The party ended up in a mangrove swamp, and some of its members were almost swallowed by the mud. Again the local Aborigines were spared.

If Tench was unenthusiastic about the whole affair, Dawes had gone further, at first actually refusing to be involved. He changed his mind only after a long talk with the Reverend Richard Johnson. But later he announced in public that *he was sorry he had been persuaded to comply with the order*. This apparently infuriated Phillip, particularly after Dawes refused to retract, and might have contributed to Dawes' ejection from the colony the following year. It is not clear why Phillip was so uncharacteristically violent in this matter. He had been sick for some years, and while the nature of his illness is unknown, it could have involved calcium deposits in his urinary tract, a common problem in sailors. It has been suggested the prevalence of this painful condition helps explain the notorious irascibility of many naval officers of the time.

In Batavia the *Supply* had contracted a Dutch ship to bring provisions to Sydney, and the *Waaksamheyd* finally arrived on 17 December 1790. At this point the immediate risk of starvation ended. The ship carried a letter from a Dutch official to Governor Phillip about *something of a Spanish war having been declared against England in May*, wrote Elizabeth, but *the particulars are not … well understood, as the letter is written in Dutch, and no one here understands enough of the language to transcribe it correctly*.

The fate of the first Norfolk Islanders continued to occupy Elizabeth's imagination. To her old friend Bridget Kingdon back in Bridgerule she wrote that in contemplating the situation of the islanders *a chill seems to overpower my faculties; my mind has so truly entered into their distresses that a dread comes over me which I am unable to describe*. Her concerns were calmed when the *Supply* was sent over and returned to say that all was well. In fact, the islanders were better fed than themselves, due to the discovery of a population of a plentiful and nutritious species of sea bird, which they had proceeded to eat. Elizabeth sensed here *so firm a reliance on the merciful dispensations of an Almighty, whose hand I think we may here trace without presumption*. Modern readers may have difficulty taking such a sentiment seriously, but Elizabeth Macarthur believed it quite literally. Such beliefs had helped her risk her life to come halfway around the world. Of course there are many Christians today who would say the same thing, but most of them also rely on life insurance and seatbelts and the welfare state and inoculations. Elizabeth did her best to minimise the risks of life too, but often there was little she could do except hope and pray; compared with our own, her life was lived in the full blast of unpredictable, and relatively unavoidable, dangers.

John became seriously ill again. The drought continued, and there was no more than a week of rain in their first nine months, much of it in the form of violent thunderstorms that broke periods of intense heat brought on by scalding winds from the north. The temperature soared to 104 degrees Fahrenheit in December 1790, when John's health recovered.

1791

In January the Macarthurs moved into a new house which contained a pianoforte belonging to Surgeon George Worgan of the *Sirius*. It is fascinating to reflect for a moment that, among the life-saving stores and essential goods packed into the First Fleet, space had been found for a piano, even if only a small model. How appealing it must have looked, sitting probably on a beaten earth floor – maybe on a rug of some kind – in a wattle-and-daub hut. Worgan, who was to depart for England shortly, shared Elizabeth's enthusiasm for landscape and gave her piano lessons. She learned to bang out *God Save the King*. The surgeon, who was kind enough to tell her she had acquired the ability to read music *with great facility*, offered to leave his pianoforte with her.

Outside, the landscape was burned brown and yellow, and Elizabeth was unimpressed, although perceptive enough to notice that whereas the settlers' gardens were barren, the native plants were *flourishing even to luxuriance*, despite the heat. In her letters we see her moving between a disdain for the obvious shortcomings of the country and an impressive aesthetic sympathy. Even though the soil was *most wretched*, the native flora contained *fine shrubs, trees, and flowers, which, by their lively tints, afford a most agreeable landscape*. Despite its drawbacks, the landscape *had symmetry of form also to recommend it, as the ground, in all the parts that have been discovered, is charmingly turn'd and diversified by agreeable vallies and gently rising hills*. The harbour's points and islands were *romantic*. She was open to the new country in a way some of the others, such as the acerbic moralist William Hill, were not. Crucial to such openness was an intellectual curiosity in subjects like astronomy, geology and botany. Such interests, widespread among the British as a result of the scientific revolution, began as amateur activities but often developed into more serious pursuits. Botany, of course, was the most important. The first years of the colony were inevitably frustrating and unproductive because of the need to learn about how the soil and climate treated crops, and about the uses to which the local flora could be put. But gradually trees were found that provided useable timber, and crops were drawn from the earth.

By March 1791, almost nine months after arriving, Elizabeth, due to John's illness and the heat, had been no further than three miles from Sydney Cove, apart from one expedition to Watson's Bay, and another to what is now called Randwick, to look out on the famous Botany Bay in the distance. She was always accompanied by at least one soldier on these walks in case of problems with the natives, but in fact this was unlikely. The British had been fortunate with the Aboriginal people of Sydney. Before leaving England, Phillip had thought he might need to build a fort and establish the town behind a river, or even on an island, the better to ward off attacks. In the event, the Aborigines were just not that warlike, and in any case their numbers were soon cut down by the smallpox. This absence of resistance was very important in the early success and spread of Sydney. (In retrospect it, perhaps only superficially, gives it a moral quality different from that of the more brutal dispossessions that were to follow.) The surviving Aborigines began to come into the town frequently, dining with the officers, and some even lived there. One Aboriginal woman known to Elizabeth brought her new-born baby wrapped in soft bark, and Elizabeth ordered something for the mother to eat and *had her taken proper care of* afterwards. The unnamed woman became a regular visitor.

John was getting on well now with Captain Nicholas Nepean, and the Macarthurs dined often with the governor. The entire colony was on short rations. In April a government farm was commenced at Toongabbie, but success continued to be elusive. Still the *Gorgon* with her officers' wives refused to appear, and Elizabeth became depressed when Captain John Hunter and the other gentlemen from the sunken *Sirius* returned to England on the *Waaksamheyd*: *In so small a society we sensibly feel the loss of every member.* The drought continued, so bad that the Macarthurs did not even try to make a garden in the small amount of land allocated to them. And yet, curiously, in the same letter in which Elizabeth told her friend Bridget Kingdon that her spirits were low, she announced that *I never was more sincerely happy than at this time.* John, Edward and she were in good health, which fact alone must have been cause for celebration for this family which had already suffered so much illness and buried one of its members at sea.

There was, of course, a society growing up around the Macarthurs. It was a government autocracy ruled by Governor Phillip, who used his power to encourage certain ways of behaving. It is often portrayed as mainly a vast prison, but it was more like an enormous government department with porous boundaries to a growing private sector. Phillip encouraged people to fend for themselves by allocating them areas of the harbour for fishing, and plots of land for farming, allowing them to keep the results of their labours (although a proportion of all fish caught had to be given to the Aborigines). He encouraged convicts to marry and have children, and freed the deserving to take up land and cultivate it.

The best known example of this treatment involved James and Elizabeth Ruse. James was a Cornish agricultural labourer convicted of theft, whose sentence expired soon after arriving in Sydney. He behaved well and Phillip gave him land and assistance so that he and his wife in February 1791 became the first settlers to be independent of the government store, much to Phillip's delight. It was something of a milestone.

Phillip imposed a degree of egalitarianism on the colony. The historian Alan Frost has pointed out how relatively free of social distinctions early New South Wales was, given the era and the sort of people there. This had begun on the way out in the First Fleet. Chastising an officer who had beaten a sailor, Phillip called all the officers to his cabin and told them that the sailors *are all we have to depend upon, and if we abuse those men … the convicts will rise and massacre us all. Those men are our support … if they are ill-treated by the officers, what support can you expect of them?* His motivation here might have been self-preservation rather than humanitarianism, but the effect was the same. After the fleet reached Sydney and food began to run short, Phillip insisted that everyone in the colony – from himself down to the convicts – be provided with exactly the same rations, an extraordinary decision for the late eighteenth century, and one not appreciated by all his officers. When the first buildings were rising at Parramatta, Phillip slept on the floor of his former servant's hut for a while. He allowed Aborigines to wander in and out of his house and shared his meals with them. For a man in his position, he was an unusually modest and tolerant human being.

Meanwhile, in Britain the many negative reports from the colony were seized upon by the evangelical Christians and others who opposed transportation. Prominent among them was Jeremy Bentham, an atheist reformer who invented the words 'codify', 'maximise', and 'panopticon', the last being a term for a circular prison he had designed himself. When democratic Britain rejected his urging for a neat criminal code to replace the messy common law, he touted his ideas about the monarchies of Europe, believing that an *enlightened despot* might receive them favourably. He was not the last reformer who would despair of democracy and hanker after autocratic powers of persuasion. For a while he had hopes of Catherine the Great of Russia. Later it was the dictator Napoleon Bonaparte who introduced a criminal code into France, although Bentham cannot be blamed for that. In 1791 Bentham wrote a pamphlet urging the British government to stop transportation and build prisons at home, and criticised New South Wales as a disaster in the making. The anti-convict movement was slowly, very slowly, gathering steam, although as Britain still supported slavery, the number of people concerned about convicts remained small.

In June 1791 the Macarthurs moved inland with Captain Nepean. Due to the gradual success of the farming carried out there, Parramatta – the original name, Rose Hill, was replaced by the Aboriginal one very early – had become the main settlement of the colony. Indeed, in June 1790 Phillip actually said that if he'd had more time to explore on his arrival in New South Wales, he would have selected Parramatta as the capital from the beginning. He had even considered this when it was discovered, three months after the First Fleet arrived, but *it required some little time to do away with the general opinion that such a situation could not be healthy, and which I was inclined to think myself until I had examined the country for some miles round, and was satisfied that there was a free circulation of air.* By this point Sydney was sufficiently established for the idea to be no longer feasible. Even so, Phillip made ambitious plans for the layout of Parramatta, and lost interest in town planning for Sydney. One reason Sydney is so meanly set out today is that it was overlooked while Parramatta developed during the crucial growth years. By the time later governors realised that Sydney would be the main centre – due to the expansion of the colony and the development of the wool-export industry – it had become a planning nightmare. But that was many years later. For now, Parramatta was the centre of activity; it was the agricultural area, and being on the river was easily accessible from Port Jackson. The boat journey took half a day. In what was the first stumbling move from prison to nation, Phillip was actively encouraging convicts whose sentences had expired, as well as ex-seamen and ex-marines, to take up agriculture there, and on Norfolk Island.

Neither the marines nor officers in the New South Wales Corps received any such help, leading Elizabeth to note that *Officers have so little encouragement in this respect, that few will in future attempt it, as evident impediments are thrown in the way to check their undertaking it.* It was probably galling for an ambitious man such as John Macarthur to see land being acquired free by former criminals. It was not for this that he had crossed the world. A year before, his colleague William Hill, in the same letter in which he had blasted the management of the Second Fleet, had complained about the governor's refusal to give the officers land: *In America the officers and settlers had grants of land in proportion to their rank,* he wrote, *but those of the marines who are now here, and have borne every hardship, have no such thing, neither is there an intention of giving each their portion. In my humble opinion nothing can be more impolitic. Industry is the first essential to the welfare of any kingdom, consequently all measures that are adopted to promote it are highly commendable; and I am well persuaded that Britain will not thank our governor for acting, not only on a mean, but on an unstable plan, to the great disquiet of every individual in the colony, and the certainty of bringing an endless burden on the mother country.*

This letter is one of the first salvos in a battle over the relative virtues of public and private activity that was to continue in Australia for some time. As far back as March 1788, the marines officers had asked Phillip for land and been refused. But in this the governor was carrying out government instructions. As the time grew near for the marines to leave, London instructed Phillip to offer land to non-commissioned officers and men, but not to the officers.

Tench had visited Parramatta in November 1790, and noted that its main street was a mile long, running from the river to Government House up on the hill. It contained thirty-two complete houses, built of sticks and clay ('wattle-and-daub') and thatched. Each was 24 x 12 feet in size and housed up to fourteen male convicts. There were more buildings in side streets, for female convicts, and huts for *convict families of good character*. There were also the barracks, the store, a barn, a bakehouse, brick-kilns, a granary, a blacksmith's shop, a stockyard, and *a most wretched hospital, totally destitute of every convenience*.

The spring of 1791 brought dramatic changes to Sydney. First, the drought broke and the harvest was a success. At the end of the year at Parramatta there were over 400 acres of government land and 200 acres of private land producing crops. The second great change was the arrival of the Third Fleet and, in late September, finally, the *Gorgon*, carrying more members of the New South Wales Corps. This brought closer the day when the marines would leave.

In early November the Macarthurs returned to Sydney, John having the command of a detachment of sixty men. He was hoping to hear that he had been made a captain: in August John Harris had noted that *Mac was makint brags one day about his Captain ship*. Elizabeth preferred Sydney due to the access to ships and the company they brought. To her great joy, several of the newly arrived officers had their wives with them, as had the ship's captain and the government agent. As a result, *our little circle has been of late quite brilliant; we are constantly making little parties in boats up and down the various inlets of the harbour ... There are so many ladies in the regiment that I am not likely to feel the want of female society as I first did.*

From now the memory of the bad days of the first four years began to slip away. More ships arrived and those on them were often impressed by what they found. Philip Gidley King, on his way to govern Norfolk Island, observed of Sydney and Parramatta that *a vast tract of land* had been cleared. Four whaling vessels (probably American) were moored in Port Jackson. One day Elizabeth and a group of the newly arrived officers' wives sailed to Manly to meet up with a hunting party that had been to Broken Bay in search of ducks. The whole group consisted of the Macarthurs, the Reverend Johnson and his wife, and the Kings. On the way back, the mast of their boat gave way, but they reached home in safety.

Of course, the main activity of the colony continued to be the guarding and support of the convicts. On 18 November Governor Phillip, in a letter to Evan Nepean, noted that the Irish convicts were causing problems due to their belief that China lay about 150 miles away, and could be reached on foot. Fifteen or twenty Irish convicts had taken to the woods, *and though several of them have been brought in when so reduced that they could not have lived a second day, if they had not been found, some of those very men have absconded a second time, and must perish.* On 12 December the testy Major Robert Ross, just back from Norfolk Island, fought a duel with the testy Captain William Hill. Two shots were fired, neither man was hit, and honour was satisfied. Earlier in the year honour had not been satisfied when Ensign Duberley of the New South Wales Corps had failed to challenge another officer after he had been insulted. He was cashiered. Now Ross and most of the marines left for England on the *Gorgon*. One who stayed behind, transferring into the corps, was Captain George Johnston.

1792, Sydney

In February 1792 Major Francis Grose, commander of the corps, finally arrived with most of the rest of his troops on the *Pitt*. He had had a good trip, apart from a fever that affected some of the soldiers in Rio de Janeiro. An amiable soul, he had encouraged their convalescence with port wine. John Macarthur was the only officer of the corps at Port Jackson when Grose arrived, and probably had himself rowed out to meet his commanding officer and his new colleagues.

In April Grose wrote to Evan Nepean that *to my great astonishment, instead of the rock I had expected to see I find myself surrounded with gardens that flourish and produce fruit of every description. … I live in as good a house as I wish for.* He continued, in words that doubtless would have made Nepean happy, *since I have left England I have seen no place I like better than this.* Historian Anne-Maree Whitaker has observed that, as most of Grose's fellow settlers were starving due to renewed food shortages, it's likely this letter was intended as a contribution to British domestic politics rather than a statement of fact. It was written to support the Pitt government and Grose's patron, Lord Mulgrave, who were being criticised for their decision to establish the colony. The letter was duly released to the press.

The New South Wales Corps numbered 319, a third more than the marines detachment it had replaced. Recruiting had been difficult and Grose had been forced to take some of his men from deserters in London's Savoy Military Prison. Many of the corps were the sort of men the Duke of Wellington was soon to describe *as the scum of the earth*, but there is no evidence they were below average by the standards of the British army, where conditions for the ordinary soldiers were so miserable that only the most desperate men joined up.

The *Pitt* stayed in Sydney until April, when it departed on a return voyage to Bengal, its captain provided with £1,440 of paymaster's bills by ten of the New South Wales Corps officers. He was to buy various goods for the officers and bring them back to Port Jackson. This use of the corps' credit to speculate marked the beginning of the officers' trading activities. However, at least at the start it was not simply a matter of speculation: the colony was desperate for basic materials such as shoes and tools. Despite his strengths in some areas, Governor Phillip did not always manage his resources well. He was overly reluctant to buy in supplies following the initial agricultural failures, the drought and the *Guardian* and *Sirius* disasters. As a result of this lack of initiative, the convicts suffered considerably, some of them dying, and the free settlers, including the soldiers, had a wretched time of it. Major Grose clearly had his own ideas about what needed to be done, at least where his own people were concerned.

London

In June 1792 Captain Trail of the *Neptune* was put on trial in London for murder. The case was well publicised in Britain, and would have been eagerly discussed by many of the inhabitants of Sydney when they received the news. Although there had been criticism in the press of the conditions of the Second Fleet, Trail came to trial only because of the actions of an attorney named Thomas Evans, a sort of early human rights activist. He assembled the evidence of a large number of sailors critical of many of Trail's actions on the *Neptune*, from starving them to killing one of their number after the ship had left Sydney. Then he mounted a public relations campaign to rouse popular indignation. Trail and his employers fought back with their own campaign: in February 1792 a defence of their actions was tabled in parliament by a politician who was on-side. Evans selected just one event from Trail's record – the killing of the ship's cook – and continued to push the government, until they gave in and the trial went ahead. Trail and his chief mate were charged with murder, found not guilty, and Evans disappeared from history, possibly struck off the rolls following criticism from the judge. It was not an era kind to crusading lawyers. Trail, in contrast, rejoined the Royal Navy, dabbled in the slave trade during a posting at Cape Town, and eventually died a wealthy man.

Parramatta

In June 1792 Macarthur was sent back to Parramatta with Elizabeth and their two children. (Elizabeth junior had just been born.) Captain Joseph Foveaux had established a regular market there for the sale of farm produce and imported goods. Richard Atkins, a well-connected former soldier who had come to New South Wales to escape his creditors, was the local magistrate. A visitor to the colony, George Thompson, left some relevant descriptions. Most of the convicts were at the government farm at Toongabbie. There they cleared land and tended the corn crop. *From the heat of the sun [this was in May], the short allowance of provision, and the ill-treatment they receive from a set of merciless wretches (most of them of their own description) who are their superintendents, their lives are rendered truly miserable.* Those women who were *not fortunate enough to be selected for wives (which every officer, settler, and soldier is entitled to, and few are without) are made hutkeepers; those who are not dignified with this office are set to make shirts, frocks, trousers, &c., for the men, at a certain number per day; occasionally to pick grass in the fields, and for a very slight offence are kept constantly at work the same as the men.* The men, for very trifling offences, were *taken to a cart-wheel to receive a Botany Bay dozen, which is 25 lashes.*

The convicts at Toongabbie were starving, Thompson had heard, and deaths were common.

Not far away lay Parramatta, *the grand settlement*, with a large hospital, church and storehouses. Government House was on Rose Hill, *a most delightful spot*. Here was Cumberland Park, where the government cattle grazed. Down the river lay Sydney, the main base of the governor and the chief officers. It contained only enough convicts for the needs of the storehouses, fishing boats, and the officers and their gardens. Ex-convicts who wanted to return to Britain also congregated here, although few of them had made it back. The colony had bricks but still no lime for mortar, so that it was impossible (or at least unwise) to erect buildings of more than one level. *There is not such a thing in the colony as a set of stairs*, Thompson marvelled, *except in the governor's house, which is ... built with lime brought from England*. In Sydney Cove *your ships lay within fifty yards of the shore, in five fathoms and a half, and as smooth as in a fish-pond*. Beyond this was *the most complete harbour in the universe*, full of oysters, cockles and fish: *snappers, mullets, light horseman (so called from their head resembling much a light horseman's cap), flatheads, salmon, whitings, and there are many other kinds of fish too tedious to mention*. And, of course, the occasional whale.

In early October Major Grose wrote to the governor explaining that the officers of his corps had decided to hire the ship *Britannia*, then in port, to fetch supplies. Grose asked Phillip to ensure it was not interfered with. This referred to the problem that it was illegal for any ship to trade within the area controlled by the East India Company, which had a monopoly on trade in the Indian Ocean and Asia. The motivation for this hiring, wrote Grose, was to relieve the suffering of the soldiers, who *at this time have scarcely shoes to their feet, and who have no other comforts than the reduced and unwholesome rations served out from the stores*. Phillip replied that he had no influence over the East India Company. He regretted the inadequacy of the stores – *which I should gladly do away by any means in my power, yet I cannot acquiesce with you in thinking that the ration served from the public stores is unwholesome; I see it daily at my own table*. Clearly the officers of the army were used to higher living standards than were to be found in the navy. Phillip confessed himself perplexed by the officers' desire to charter a ship, but finally bowed out of the matter, over which ultimately he had no control. The corps was financially independent of the governor. In this case we see more evidence of Phillips' asceticism, which was still being demonstrated in his persistence in giving the same rations to everyone in the colony, which was provocative. Certainly, Grose was provoked, protesting at this time to the Deputy Secretary of War and informing him indignantly how in Sydney *the captain of a company and the convict transported for life divide in share and share alike whatever is served out*.

The investment in the *Britannia*, for £4,000, indicated that the New South Wales Corps' officers were becoming more serious about trade, and represented an acceleration of the growth in complexity of the colonial society. The officers were to charter the *Britannia* again, and later in the decade they hired another ship; however, far more common would be their practice of banding together and buying all or most of the stock from incoming ships. These trading activities meant that the officers, already dominant in the provision of violence and justice in the colony, were now expanding its economy and taking a dominant position there too. The potential for conflict of interest was growing by the month.

Yet the officers did not trade alone. In October 1792 the *Royal Admiral* arrived from London and its captain proceeded to sell alcohol in shops set up in Sydney and Parramatta.

Scenes of public drunkenness occurred that appalled David Collins. *Several of the settlers … conducted themselves with the greatest impropriety, beating their wives, destroying their crops in the grounds and destroying each other's property*. The corps bought some kegs of spirits for its own use (a daily ration of spirits was normal in the army), and a dispute arose when one of these was landed from the ship without the appropriate regulations being followed. A wharf constable protested and was ignored by the officer involved. Somehow, John Macarthur was given the job of explaining things to Governor Phillip, who resented his tone and threatened him with arrest. *Sir*, replied Macarthur, *you may please yourself. You are the first officer that ever threatened me with arrest. And I give you my word of honour if I am put in arrest, I shall require a full and sufficient explanation of the cause before I consent to sit quietly down under such a disgrace.* He then left, and refused all further invitations from Phillip to dine at Government House. This did not stop Phillip presently giving him a fourteen-year lease on the property the Macarthur family occupied in High Street (later George Street), one of the first leases to be granted in Sydney town. The land measured 100 by 200 feet. Elizabeth noted that *We are at present here rather in an unsettled state, which is not very agreeable*. However, in December she was also able to report that *Edward grows a strong, healthy child, and from being a great deal of trouble to me ceases to be almost any at all. He prattles everything, and is quite papa's darling*.

On the same day that Phillip queried the hiring of the *Britannia*, he had written to London asking for instructions on how much land he might give the officers of the corps, but only *on such officers becoming a settler*. He pointed out that the officers had been led to believe, before coming to Sydney, that they would receive land and convicts to work it. Although *very far from wishing to throw the smallest obstacle in the way of offi-*

cers obtaining grants of land, he was concerned that giving convicts to officers would detract from public works and government agriculture. Major Grose wrote to London requesting land for his officers, *absolutely the only chance of comfort that I know of.* He believed that public agriculture had failed to feed the colony adequately, and that some private enterprise was urgently needed.

While Phillip did not want officers to have land unless they resigned their commissions, Grose believed a man could be an officer and a farmer at the same time. This, like some of the earlier altercations between Phillip and Ross, is indicative of the troubled relationship between all the naval governors and the military. Its roots lay partly in the different officer cultures of the navy and the army. If one considers the position of the naval captain it is apparent that it was relatively solitary, authoritarian, intellectual and ascetic. There was a body of learning needed to navigate a ship. Harsh punishments were required to ensure that rules were followed promptly and to the letter. Often a ship's commander was alone, faced with decisions, ranging from the diplomatic to the military, that might affect Britain's interests. There was also something monk-like about the lives of most naval officers at sea, with the rough conditions, bad food, prolonged periods of celibacy, and few opportunities to submit to temptation of most sorts except verbal and sometimes physical violence towards subordinates. This is the culture the governors came from: professional, self-denying, and harsh. In comparison, the army officers led easier lives, ones that kept them in touch far more with the society around them. The job was simple enough, requiring no skills beyond those possessed by most country gentlemen. For them, rank was more likely to be a way of obtaining comforts, and of bettering oneself.

As a result of these differing backgrounds, the sailor and

the soldier often saw service in New South Wales in profoundly different terms. This was rarely stated – often the most important things are not – but it lay beneath many of the squabbles that were to be a feature of the colony's history for years, as the government persisted in appointing naval governors. Another difference was to be found in the soldier's and sailor's experience of private enterprise. Military officers often engaged in commercial activities, sometimes using the army's funds. Naval officers' only connection with the economy – apart from the rare sale of a prize vessel – was in being ripped off by contractors and other merchants. At this point New South Wales already had a primitive economy at all levels. Many convicts brought money and goods with them from Britain, which they used to trade for stolen food and other items. Female convicts sold their bodies to the sailors and soldiers on the way out. In Sydney, free men and criminals collected seeds that they sold to sailors to take back to the nurseries in England. These were just some of the small commercial activities that occurred from Sydney's first day. Many of the army officers felt quite at home with financial matters and wanted a piece of the action.

This continual conflict between the two cultures wore down all the naval governors. Phillip was sick and had asked London several times for permission to go home. It was never explicitly given, but by December 1792 he had had enough. The colony was not yet self-sufficient, but it had been a good year, despite some hunger. The amount of government land under cultivation had increased from 400 acres to over 1,500 acres. There were now 72 farmers in the colony, all smallholders, mostly ex-convicts and ex-marines. There were already 688 sheep. So Phillip decided to go. The New South Wales Corps saw him off, and on 11 December the good governor passed through the heads on the *Atlantic*, never to return. The comment that perhaps comes closest to revealing his inner self had been published in his 1789 book about the voyage to Botany Bay: *There are few things more pleasing than the contemplation of order and useful arrangement, arising gradually out of tumult and confusion; and perhaps this satisfaction cannot anywhere be more fully enjoyed than where a settlement of civilised people is fixing itself upon a newly discovered or savage coast.*

The hour had called forth the man: in the past five years Phillip had done more than most could have. But it was the right time to leave. To turn what was essentially still a government department into a free society, other, perhaps less obviously admirable, men would be needed. As would a descent into tumult and confusion.

III

Elizabeth Farm

Major Francis Grose was about thirty-four when, on Phillip's departure, he became acting governor of New South Wales in December 1792. He enjoyed his comforts and had a well-developed notion of the respect due to himself and his regiment. On seeing the accommodation proposed for him on the *Gorgon* back in England, he had retreated to the Salopian Coffee House and written to Evan Nepean that, although *It is with much reluctance I complain*, the quarters *are not such as we consider ourselves entitled to*. Indeed, *I am the first officer of my rank who sailed with so few comforts*. In the event, he did not sail on the *Gorgon*, which suggests his protests were more effective than Macarthur's had been on the *Neptune*.

Nothing in Grose's background, beyond normal service in the army, had prepared him for the role of governor. But there is no reason to believe he was daunted, and he was soon to show a considerable capacity for delegation, and some new ideas on how the colony ought to be developed. One of his first actions was to increase the power of the military officers at the expense of the civilian ones. He then upset the Reverend Johnson by reducing the number of religious services from two to one a day. Johnson pointed out that not everyone from each residence could come to the same service, as some had to stay behind to prevent their homes from being robbed while the others were absent. But his argument was ignored. Grose also cut the services short, and Johnson complained that one morning *at the beat of a drum, all soldiers instantly rose up, took their arms, and marched away. I had then been just three quarters of an hour from the beginning of the service.* The overseers were supposed to bring all the convicts to services, but Johnson said that only ten in a thousand turned up.

As acting governor, Grose had the power to issue treasury bills to anyone he liked. These could be drawn on the British government and were therefore as good as cash. On 9 January 1793 he wrote to Henry Dundas, Principal Secretary of State at the Home Office, to announce that he had purchased supplies from the *Hope*, an American sealing vessel that had called at Port Jackson. *I lamented on this occasion being obliged to purchase his spirits, without which he would not agree to the disposal of his provisions*, Grose told Dundas. In the same despatch, Grose noted he was much plagued with ex-convict settlers who wanted to sell their land and livestock so they could return to England. Phillip had kindly distributed a flock of sheep among these people on his departure, but they had promptly offered them for sale. *I was absolutely obliged*, wrote Grose, *to encourage and promote the purchase of them by the officers, dreading that, without this precaution*, the sheep would soon be killed and eaten, thereby destroying breeding stock that had taken years to assemble.

Grose proceeded to sell *a considerable quantity* of the spirits from the *Hope* to the civil and military officers for £405. According to David Collins, some of this found its way into the hands of convict superintendents, and from there into the mouths of some of the convicts themselves, upon which terrible scenes ensued. *A discreet use of it being wholly out of the question with those people, intoxication was become common among them.* Collins noted that their fondness for spirits was *incredible*, and they *preferred receiving liquor for labour, to every other article of provisions or clothing that could be offered them.* The convicts, and indeed some of the small settlers, had had almost no alcohol for a long time and some of them now proceeded to get plastered. Collins recorded the fate of a small boat of settlers who, after *drinking and revelling* in Sydney, were making their way up the river to Parramatta. The boat tipped over due to the clumsiness of one of the group, and the wife and child of Charles Williams were drowned, along with another woman. Williams and his wife, said Collins, *were noted in the colony for the general immorality of their conduct.* Williams buried his wife, who had been six months pregnant, near his front door, and presently *he was seen sitting at his door, with a bottle of rum in his hand, and actually drinking one glass and pouring another on her grave until it was emptied, prefacing every libation by declaring how well she had loved it during her life. He appeared to be in a state not far from insanity.*

On 16 February Grose told Dundas he had changed Phillip's rationing system, so that the soldiers now received more than the convicts. He also appointed John Macarthur to a new position, inspector of public works at Parramatta, which *I am unequal to manage by myself.* The position was unpaid, as was the position of magistrate that many officers were to fill in the coming years (although sometimes extra convicts were

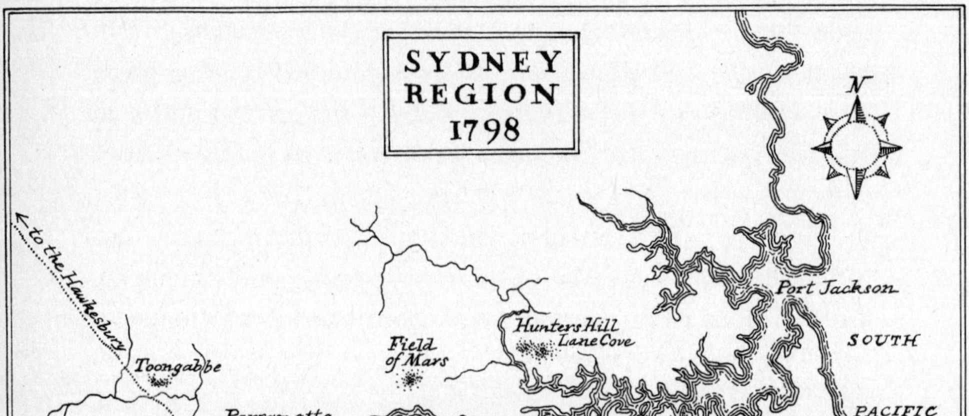
SYDNEY REGION 1798
N
to the Hawkesbury
Toongabbe
Parramatta
Prospect Hill
Elizabeth Farm
Field of Mars
Hunters Hill
Lane Cove
Field of Concord
Sydney-Parramatta Road
Petersham
Port Jackson
Wooloomooloo
Sydney
SOUTH PACIFIC OCEAN
Cooks River
Botany Bay
Bankstown
Georges River

allocated). Gentlemen were expected to serve society in this manner. Macarthur now ran the day-to-day affairs of government at Parramatta and Toongabbie. He was responsible for the government farms and the public stores of grain, and the allocation of convicts to their jobs. *In deputing a trust of this kind it becomes an object of some importance to repose it where there can be no doubt of its being executed with the strictest attention … I am confident my judgement will not be discredited in the choice I have made.* Grose could report that already *we get a great deal more done than we used to do, and that the work in general is much better done.* It's interesting that Macarthur was chosen rather than a more senior officer, such as Nepean. Grose must have been impressed with him. It's also possible the extra work involved, and perhaps the need to spend so much time at Parramatta, had no appeal for the others, whereas Macarthur saw the job as an opportunity for advancement.

A new 60-acre public farm was cleared in Petersham this year, and Macarthur supervised the change from large plantings of maize to wheat. Maize grew better in dry conditions, but was not a central part of British cuisine and most settlers never grew to like it. He continued to build up the numbers of government livestock, none of which had ever been killed for meat. Most of the stock brought by Phillip from the Cape of Good Hope had died on the way, and the remainder – all but one sheep – had escaped in July 1788 due to the inattentiveness of their convict overseer. More were brought from the Cape later, but it was an expensive business. The inadequate number of cattle was to be a problem for many years.

Finally, in this despatch which announced the introduction of so much important change into both the colony and the life of John and Elizabeth Macarthur, Grose revealed that he had allotted land to those officers who wanted it. A letter had arrived from Henry Dundas to Phillip, giving the governor permission to make such grants. Macarthur was given 100 acres on the south side of the creek leading to Parramatta. It was some of the best land yet found, next to James Ruse's successful property, and would soon receive the name Elizabeth Farm. The reason why the early grants were relatively small was that experience in Canada had taught British administrators that overly generous grants were often left unfarmed for years. Grose gave land to officers at Parramatta, Petersham and around the harbour, and to the ordinary soldiers at Concord and Hunters Hill. Grants were also made at Georges River and Botany Bay, and to emancipists and free settlers at Liberty Plains, Concord, Petersham, the North Shore and Toongabbie.

And so, some three and a half years after John and Elizabeth had decided to cross the world in search of *material advantages*, they finally received something. It had been a long time coming. They had risked their lives, and that of their boy and their maid, had lost a baby, John had fought a duel and acquired a recurring illness, and they had lived for several years in isolated, boring, difficult and uncertain circumstances. There is no record of their feelings on being told of Grose's decision, but we can imagine they must have been ecstatic. God had not abandoned them after all.

In Grose's letter there is a sense of energy unleashed. He added that the officers were clearing their land already and *Whether their efforts result from the novelty of the business, or the advantages they promise themselves, I cannot say, but their exertions are really astonishing*. Grose predicted that the officers were the only *settlers on whom reliance can be placed*. Many historians have written off Grose as a leader simply out to enrich his officers. But it's more complicated than that, and, whatever his motivations, there were arguments for what he did. The colony needed food for its commissariat, the government store that fed virtually everyone. Government farming, like most activities conducted for the common good, was not very productive. Nor were the ex-convicts, in Grose's view, successful farmers. So he was proposing the officers as an alternative. He also gave land to a small number of recently arrived free settlers and to the civil officers, such as the Reverend Richard Johnson and Commissary John Palmer. The grants made to the officers in New South Wales were unusual – elsewhere in the empire, soldiers generally could not receive land unless they resigned their commissions.

These land grants were a defining moment for the colony, as it split the officers' loyalties between public duty and private desires and hastened the establishment of a private sector outside the closed system run by the governor. Some observers have been quick to point to the ill effects which were to flow from the officers' increasing involvement in commercial activities, but this also gave a tremendous fillip to the colony's economy.

Grose told Dundas the officers were clearing their new land *at their own expense*. In fact, each had been given, in addition to two convicts for domestic service, ten more, clothed and fed from the public store, to work his land. Land-owners were also able to pay other convicts to work for them after hours or on their days off. The Macarthurs pressed ahead, clearing land and beginning to raise a house. The generous Grose gave them what Elizabeth described as *to a family in this country in its present situation … a gift beyond any value that can be placed upon it*: a cow in calf. The family still lived in Sydney, although Macarthur was spending most of his time at Parramatta.

Also in February, the *Shah Hormuzear* arrived from India with about a hundred so-called Bengal sheep on board. This has been called the foundation flock of Australia, as before its arrival there was only a small number of Cape sheep in the colony. Both breeds were quite hairy. Macarthur acquired thirty ewes, probably later in the year from Captain Nicholas Nepean when he left the colony. Ten years in the future he was to say that half his flock, which by then numbered thousands, was descended from these sheep. The Bengals were prolific, breeding twice in thirteen months and often giving four lambs at a time. Macarthur had not been among the officers who had subscribed to the voyage of the *Shah Hormuzear*, possibly because at that stage he could not afford to. Captain William Hill was among the subscribers, but presently sold his sheep to Captain Joseph Foveaux and left the colony. In Torres Strait he went ashore with a party looking for water and was killed by natives.

On 13 March 1793 two Spanish ships, the *Descuvierta* and *Atrevida*, arrived in Port Jackson. The opinions of their officers regarding the colony are of interest. On his return home Alexandro Malaspina, the expedition leader, reported that Port Jackson was a disaster that would drain Britain financially; or, as he put it, referring to an explosive device, a *mine already opened for the not distant ruin of this ambitious power.* He had arrived expecting to view the results of rapid progress, but found at Sydney *land ill-cleared, fields little worked, wretched houses, and everywhere the marks of oppression and disgust.* At this time Sydney contained some *300 houses, the greater part constructed and roofed with brick and tile (a third part still with straw), gathered together, it would appear, in disorder.* At Toongabbie, now managed by Macarthur, the convicts were being fed with *a miserable ration so putrid that its odour was not supportable at a great distance.*

The Spaniards were fascinated by the British character, *prone to vehement passions ... Their swordsmen, their frequent, at times baseless duels, the continued abuse of liquor.* There was no religion apart from the philosophy of liberalism, described as *an absolute liberty of manners insofar as does not harm the established public order.* All the women in the colony apart from the officers' wives were prostitutes, who almost always *approach announcing the price at which they sell their favours, and it is common to claim a small increase to indulge other detestable vices apparently frequent in the colony.*

This somewhat baroque account gives some glimpses of early Sydney not provided in the English descriptions, especially where sex was concerned, and indicates the gulf that existed between the two European cultures. For instance, the British took a keen if patronising interest in the Aborigines, but for the Spaniards they were *the most miserable and least advanced nation which exists on earth ... they appear to occupy the last grade of man before passing on to the ape family.* Aboriginal women too were keen prostitutes, although *it would require a total depravity of sensibilities to suffer the smell, the slovenliness and the roughness of their gross caresses: few women in the universe could present themselves in conditions which would cause an effect so contrary to their desires. However, it happens that such is the desire for novelty and the depravity of tastes in a free country where prostitution is so common, and where nothing is as easy as the means of satisfying every kind of voluptuous passion, that the English lower orders do not disdain them, and each night a large number of them are gathered in the quarters of the troops.*

Malaspina was impressed, though, by the smallness of the colony's administration. Unlike many of the settlers, he commended the fact that the judge-advocate had no previous knowledge of the law and that the commissary ran the local economy. *Simplicity will always be the basis of good legislation in distant colonies,* he opined, contrasting the minimal British approach with the elaborate bureaucracies of his own people. Elsewhere he noted that kangaroo meat was frequently eaten.

The kangaroo was hunted with greyhounds but even so, at times the hunters had to chase it for miles before making a kill. He found the meat *very good sustenance, inferior to veal, but better than many others ... without the gameyness of goat, nor as greasy as mutton.*

Whereas Malaspina's report veers between disdain and praise, the accounts by some of the other Spaniards were generally positive. Botanist Don Luis Née was impressed by how agriculture was pushing back *the bush which disfigures the land*, and remarked how the British in Sydney had *transformed a rough and uncultivated country into a pleasant garden. It had been in existence for hardly five years, and yet had the appearance of an old establishment.* The Spaniards were struck by the solidity of the brick cottages they passed, most neatly kept and with their own orchard or garden. One at Parramatta was covered with roses.

Née, like some of the British, noted that George Barrington, once a notorious London pickpocket, was now in charge of the constables at Parramatta. The Spaniard found this funny, which is how it has often been treated by later commentators. However, it symbolised a moral and social reversal that seriously disturbed many contemporaries. As more and more convicts were freed in the coming years, ethical compromises such as Barrington's promotion become an engine pushing the social and political development of Australia away from its British roots.

In May Macarthur took over command of the Parramatta garrison from Captain Foveaux, and the next month Elizabeth and he, together with Grose and his wife, spent some days at the Lodge, the residence of Richard Atkins, the most socially significant man in the colony. He was charming and handsome, the third son of Sir William Bowyer, Baronet. The family was extremely well connected, and his oldest brother was a general and his next one an admiral. He had been left a legacy by his cousin, Sir Richard Atkins (hence his change of surname), but had run through it and amassed debts, assisted in this downward progress by a love for the bottle. Hence his appearance in Sydney in 1792, at the age of forty-seven, on the same ship as Grose. Phillip had greeted him as befitted a man with his connections, appointing him registrar of the Vice-Admiralty Court (an irrelevant post) and justice of the peace. Grose had granted him 100 acres at Parramatta, and the day after the dinner with the Macarthurs promised him ten convicts to work it. In June Dundas from the Home Office told Grose the government would pay for only two convicts per officer, for a period of two years. Grose cheerfully ignored this. Dundas also instructed that spirits not be sold to convicts.

In June the *Britannia* returned full of supplies from Cape Town to find the colony with only ten weeks' stock of food left in the store. This would seem to justify Grose's trading as an alternative to Phillip's less imaginative preference to cut rations and let people suffer. The officers did a roaring business in their badly needed goods, and sent the *Britannia* off to buy more. Macarthur played a prominent part in this trade, for the three years from 1793 keeping the records of the officers' import activities. In their trading enterprises the military were joined by some of the civilian officials, such as Commissary John Palmer. Eventually they were to employ agents in Calcutta and London to assist them in their commercial dealings. While the officers were often criticised at the time, and have been since, for these activities (particularly their monopoly of purchases from incoming vessels and the high prices they charged), two considerations help put this in context. The first is that this group was the only one with the funds to make such purchases. For capital the officers used their own money, such as their salaries and store receipts issued by the commissariat when they sold it produce from their farms. They also used the unpaid wages of their troops and other army funds, to which they had access (as was common throughout the army). Sometimes they could even borrow from the commissariat as if it were a bank. The second factor affecting any judgement of the officers is that, had they not acted as they did, the captains of visiting ships would have set up shop in Sydney and charged the high mark-up themselves, as they still did on occasion. The other colonists would have been no better off, and all the profits would have gone off shore.

The officers' trade often involved spirits, known by the generic name of rum. (Beer did not travel well, and few Britons drank wine.) This fact has been used to blacken the reputation of the corps, because they had a privileged access to the supply of spirits and charged large mark-ups. Following the so-called Rum Rebellion in 1808, Governor William Bligh described the system that had operated since the early 1790s. *It is inconceivable*, he wrote, *what money has been made by some people, through the means of bartering spiritous liquors*. As Bligh indicated, profits often came from bartering spirits rather than selling them, because there was so little cash in the colony. In a nutshell, the system worked like this. Say you wanted to obtain a pound of mutton worth eighteen pence. You could either pay in cash or barter for it with, say, wheat you had grown or spirits you had obtained that were also worth eighteen pence. However, if you were an officer, the spirits you owned that had a market value of eighteen pence might have cost you only six pence, so you had tripled your money. This was solely because you had privileged access to spirits because you had the cash or credit with which to import them, or buy them from importers.

Many critics railed against the large mark-ups involved, but the ability to mark up depended on market prices, which were determined by the amount of liquor in the colony. Later it was the governors, sometimes acting on orders from Britain, who restricted the supply, by banning stills and limiting imports. They did this because they were concerned about the evil effects of liquor on the populace. But, whatever benefit this policy might have had for the sobriety of the lower orders, its economic effects were less positive. Many ordinary people wanted to drink (as did many officers). They were far from home and usually sexually frustrated: there were about five times as many men as women in Sydney at this time, which meant that for many male convicts, transportation to Botany Bay meant they would never have sex with a woman for the rest of their lives. (Presumably the capacity of the local prostitutes, and the ability or inclination of the male convicts to take advantage of this, were limited.) Many had also been enthusiastic drinkers back home: Britons in the eighteenth century drank a lot of alcohol. So to expect those transported to Australia to change their habits was ambitious. To create a situation whereby, in order to satisfy their craving, they had to use up their savings and their assets was, on the part of government, irresponsible.

It has been claimed that the officers would buy entire land grants from wretched alcoholic emancipists for a bottle of rum, but this too appears to stem from wishful thinking – or hating – on the part of the accusers. There is little evidence of this. It's true that relatively few emancipists hung on to their land grants for very long, but in most cases this was probably because, as city folk without experience of agriculture, they had no desire to be farmers.

What really happened in Sydney in the 1790s occurred because trade is a natural human activity, and at that time and place the corps, thanks to the way the colony had been set up, populated, funded and regulated by the government, was the group in the best position to trade. The amount of profit they obtained is not known, but it was spread around, particularly among the soldiers, convicts and emancipists they used to retail the goods they sold. (Being gentlemen, most officers did not set up shop themselves.) Emancipist Simeon Lord acted as agent for Adjutant Thomas Rowley, launching himself on the path to becoming a very wealthy man. Eventually he and other emancipist traders would go into competition with the officers' monopoly and break it.

There were plenty of complaints about the traders. Robert Murry, a mate on the *Britannia*, observed that *the meannesses practised by the officers of the N.S.W. Corps exceeds all I could possibly suppose usurers capable of.* He claimed the *United Company of Traffic Merchants* had ruined one small settler by giving him spirits and then taking his farm in lieu of payment, at only one quarter of its true value. He suggested the officers of the corps were unlike any others in the British army, that some lived with prostitutes, and neglected their public duties. All this must strike a man of common feelings *with horror and detestation.* One of the officers' tricks, according to Murry, was to import Spanish dollars for use as currency, and to cut them up into small pieces and exchange them for more than they had paid for them. He wasn't sure who had invented this scam, but the rumour was that it was a certain John Macarthur.

So, for much of the 1790s, the officers made a lot of their money at the expense of the British taxpayers and their fellow settlers. They invested funds that were not their own and the risk was relatively modest. On the farming side, the commissariat purchased everything they grew using land taken from the Aborigines and the labour of convicts provided and fed by government. Free enterprise in Australia had an assisted birth, as it was frequently to have an assisted life in the years to come. The key to the officers' success was that they had been born in a prosperous nation, which continued to support New South Wales for many years with a stream of gold, a trickle of which they managed to divert into their own pockets. John Macarthur benefited from this as much as anyone, and took more advantage than most of the opportunities presented to him.

While the officers of the New South Wales Corps often fought among themselves, they tended to preserve a united front against outsiders. There were various self-interested reasons for this, not least the raw fact that they were surrounded by criminals who didn't like them. Robbery was incredibly common, and people were afraid. Richard Atkins was woken one night by a noise in his storeroom. When he went in with his sword he found a keg of wine and a bag of wheat had gone, although he discovered them the next morning in his yard. He noted in his diary that *the last two ships from Ireland brought us upwards of 200 of the greatest villains on earth, since which times crimes and riots have multiplied*. He felt Grose was too lenient, and hoped he would soon *make an example* of some convicted criminals, because *severity well placed is mercy*.

Towards the end of 1793 the Macarthurs moved permanently to Elizabeth Farm, where their new house had been built on a hill overlooking a creek. It was a brick building, 68 feet wide and 18 feet deep, with four rooms, a large hall, and a cellar. The kitchen and servants' accommodation were to be separate buildings, giving the rear section of the house, once finished, the air of a set of pavilions. It had, like almost every other building in the colony, plenty of fireplaces and just one storey. The house was surrounded by a three-acre garden and orchard.

This building still exists today, although it has been added to and enlarged considerably. (John Macarthur was Australia's first obsessive renovator.) The Macarthurs were to live in it until almost the end of John's life. It was the dream they had left England to pursue, the embodiment of a rural vision with its roots in Devon and Cornwall and the world before the industrial revolution. In a sense, the Macarthurs were building the sort of Arcadia they had learned about from the Bible and the ancient Greek and Roman writers, and in the recent pastoral poetry of England. In 1770 Oliver Goldsmith had published his poem 'The Deserted Village', a protest against the changes that were transforming the British countryside. Now the Macarthurs had escaped to the other side of the world and were recreating that lost world. It involved a nostalgia whose expression was made possible by free labour and the British taxpayer, but was none the less powerful for that.

It also, of course, depended on the taking of land from its original possessors, the Aborigines. There is little to suggest the British settlers felt any guilt at all over this. In the late twentieth century the concept of land rights gained widespread currency, and there were many arguments over how the colonisers viewed their actions in legal terms – whether, for instance, they regarded the land as *terra nullius*. While this modern legalistic approach has helped Aboriginal people gain land and other benefits, it should not obscure the fact that the original settlers had such an enormous *philosophical* belief in the justice of what they were doing that legal concerns simply never occurred to most of them. Today we tend to see issues like this solely in terms of racism, but for the Macarthurs and their contemporaries the issue of land use was probably just as important. It was an age that believed deeply and passionately in the moral benefits of progress and improvement. No one, whatever their colour, could get in the way of development.

In 1790 Matthew Boulton, a hugely successful English manufacturer, boasted of the benefits of enclosing public land for one of his factories: *I founded my manufactory upon one of the most barren commons in England, where there existed but a few miserable huts filled with idle beggarly people, who by the help of the common land and a little thieving made shift to live without working. The scene is now entirely changed. I have employed a thousand men, women and children, in my aforesaid manufactory, for nearly thirty years past.* This was a common view: and if English people in the Midlands could be dispossessed with such righteousness, what chance had the Aborigines of New South Wales?

Colonisation and the issues it threw up required old, not new, ways of thinking and acting. It involved far more violence, brutality and risk than ordinary life back in Britain, and would have been hampered by a rigorously legalistic approach. So the flexible notion of honour, for instance, was more useful to the gentlemen colonisers than the rule of law. They needed a behavioural rule of thumb that would enable them to respond quickly to threats and opportunities. Perhaps there was some recognition of these considerations in the way the British parliament tended to leave colonisation issues to royal authority and ministers. All this helps us understand why the British government, sophisticated in many ways, was prepared to allow New South Wales to struggle with such a primitive set of legal arrangements – thereby providing circumstances in which the code of honour could flourish – for so long.

Elizabeth Farm was important because it demonstrated to everyone that John Macarthur was a real gentleman, indeed a member of the gentry. Now that he possessed acres, no one could accuse him of being merely aspirational. He had arrived. It was not just a matter of snobbery or social superiority, but of obligations and history. The gentleman class went back to the warriors of medieval England, to the knight in Chaucer, that *verray parfit gentil knight*. It ran through the courts of the intervening period, reflected in Lord Chesterfield's letters to his son in the mid-eighteenth century. This correspondence is one of the few efforts to specify the qualities of a gentleman. He had to be brave, well educated (Greek, Latin, history, political and other philosophy), virtuous and honourable, a man of his word who showed consideration for women. But it all had to be carried lightly: *Wear your learning, like your watch, in a private pocket; and do not pull it out merely to show you have one …*, Lord Chesterfield advised his bastard boy. *Perfect good-breeding is equally inconsistent with a stiff formality, an impertinent forwardness and an awkward bashfulness.* Much later, in the 1850s, W.M. Thackeray wrote that to be a gentleman was *to be honest, to be gentle, to be generous, to be wise, and, possessing all these qualities, to exercise them in the most graceful outward manner. … a loyal son, a true husband and honest father … his life [ought] to be decent – his bills paid – his tastes to be high and elegant.* There is almost nothing about money in either description, although one notes that rather a lot of it would have been necessary to attain some of the attributes (Greek, elegant tastes) listed.

There is no definitive description of what it meant to be a gentleman in Macarthur's time. This is frustrating for us, but not surprising when one considers how English the institution was. The English have never been great abstract theorists, and other forms of their culture, such as the law, the language and much of their philosophy, are noted for their empirical nature. It is a culture which, at least until fairly recently, was based on the accretion of tradition. In this it differed greatly from those European cultures, such as the French, with an enthusiasm for revisiting first principles and codification.

A great benefit of not being too specific about the requirements of being a gentleman was that they were flexible and could respond to changing conditions. In practice this meant that the class continued to expand throughout the eighteenth and nineteenth centuries, keeping large numbers of people happy and providing a porous barrier between the masses and the lower levels of the aristocracy. No doubt Macarthur's own views on the boundaries of gentleness shifted as he rose from being a penniless gentleman lieutenant to a wealthy gentleman landowner, but the point is that the system could accommodate both definitions and everything in between. It was quite different from the more rigid social categories in other European countries, where there were often no gentlemen (or anything like them) and the aristocracy was rigid and non-permeable. In Britain, much was flexible, which meant the country avoided revolution by the excluded and enabled the gentleman class to expand rapidly to fill the empire's need for more leaders, officers and businessmen. However, there was no doubt that possessing land helped greatly, which is why the most successful members of the burgeoning British financial, industrial and military classes bought country properties as soon as they could afford to. The best gentlemen came from good families, but even if a parvenu couldn't manage that, at least he might buy the house a good family had once lived in, or build something that resembled such a place.

It is of course easy to ridicule the idea of the gentleman, not least because the ideal was not always attained in life. But sometimes ideals are better than their absence. This one offered many British men a sort of secular religion, a code of values independent of the monarch and the law. As Edmund Burke noted: *A king may make a nobleman, but he cannot make a gentleman.*

Macarthur must have been the busiest man in the colony

at this time. As well as commanding the military detachment at Parramatta, he ran the public works. It's interesting to note that since May the young lieutenant and his seventy-four soldiers had been in charge of almost half the colony – including over half the convicts and most of its agricultural activities – while Grose, his four captains and several hundred soldiers lolled about Sydney Cove enjoying the harbour breezes and apparently doing not much at all. No wonder Grose flattered Macarthur by calling him *Counsellor* and referring to his *old head on young shoulders*.

1794

While the colony's economy might have been growing, its moral character remained a matter of distress to the Reverend Richard Johnson. As we have seen, he was keen not merely to instruct the convicts in good behaviour but to convert them to his own pitch of religious enthusiasm. It was not just the convicts who resisted: Phillip had urged him to restrict his sermons to *moral subjects* (i.e. to focus on behaviour rather than spirituality). Johnson was depressed by the resistance he met in Sydney, possibly the least fruitful place in the British Empire for his purposes. In 1793 the Reverend John Newton, an evangelical colleague and supporter in Britain, had assured him that *You will have the honour of being the first Apostle to the South Seas ... you are sowing in tears, but you shall reap in joy*, but his tears increased under the unsympathetic Grose. However, in March the *William* brought an assistant for Johnson, the Reverend Samuel Marsden and his wife Elizabeth. Marsden was thirty, the son of a blacksmith from an area of Yorkshire much influenced by Methodism. He had become a lay preacher and been taken up by the evangelicals, who paid for his education. He went to Cambridge, and was recommended by William Wilberforce for the newly created post in Sydney. After a brief spell on Norfolk Island he moved to Parramatta.

He was not happy with what he found there, later describing it to Governor Hunter as a scene of *everything immoral and profane*. John Macarthur, the commanding officer, had no respect for the Sabbath, which *was generally spent in riot and dissipation by the settlers, soldiers and prisoners*. One Sunday a drunk insulted Marsden after service, and the reverend asked Macarthur to lock him up. But Marsden had to report that Macarthur, regrettably, considered the request *vexatious, treated me in a manner unbecoming a gentleman, and dismissed the settler in his state of intoxication. The consequence was*, Marsden claimed, that the man was stabbed to death while drunk again a few Sundays later. The reverend hastened to the place to perform the funeral rights, *when upon my arrival I found his companions had buried him in the most beastly manner, after pronouncing the most horrid oaths, curses, and imprecations over his corpse*.

Unfortunately for Johnson and Marsden, New South Wales comprised the two least religious types of Briton: criminals and soldiers. This was the beginning of the failure of religion to take deep roots in Australia. Historian Patrick O'Farrell has suggested that another reason for this was that in Australia the state, from the earliest days, took over many of the functions of religion, such as charity and hospitals, thereby denying it some of the non-spiritual importance it had in Europe.

In April 1794 the *Daedalus* arrived with one calf, four sheep and eighty pigs. In the same month, Johnson wrote to Henry Dundas to complain about the acting governor. Johnson had a wife and two small children, but, unlike the Macarthurs, had been denied a cow and enough convicts to clear his land. He said he had only two convicts paid for by government, whereas *some have near twenty*. He was concerned about his status in the colony and how this treatment reflected on it. Before long, Grose was describing him as *one of the people called Methodists, a very troublesome, discontented character.* In August William Wilberforce wrote to Dundas about these problems, assuring him that Johnson was *one of the worthiest men breathing, the most active, the most humble, and at the same time very little acquainted with the world.* Altogether, he seems to have been a strange choice as chaplain to a convict settlement. Wilberforce's letter is one of the first instances of a major problem that the governors suffered, where their decisions were appealed against over their heads by men in the colony with powerful friends in London. But in this case the attempted intervention had no effect.

After Phillip and then Grose had refused to build a church, Johnson used £67 of his own funds to erect a temporary one himself. He then sought recompense from the government, which led to more bitterness. In November he complained to Dundas that *my station, calling, and standing in the colony entitle me to the same indulgences as are allowed to other gentlemen.* Even Johnson was not immune to questions of status. But was the status of gentleman worth seeking? A month later the Reverend Newton wrote to Johnson: *You have no reason to expect to be treated like a chaplain or a gentleman while you would not conduct yourself as other chaplains and gentlemen usually do. If you had conformed to their manners and practices they would not have respected you more in their hearts, but they might have been more civil.* The implication here is that gentlemen (and indeed some chaplains) were not good Christians, at least by the rigorous standards of the evangelicals.

One reason for Grose's seeming contempt for Johnson was that evangelicalism, because rooted in the lower class sect of Methodism, was not the religion of gentlemen – not yet, anyway. Johnson, who was a graduate and came from a social background roughly similar to the Macarthurs', clearly

George III
His reluctance to allow the American colonies to expand helped create the War of Independence. The loss of the American colonies led to the need for a new colony in New South Wales – where Governor William Bligh's meanness with land was to contribute to the rebellion of 1808.

Governor Arthur Phillip, father of Australian egalitarianism and first of the naval governors, who enjoyed nothing more than the contemplation of order and useful arrangement, arising gradually out of tumult and confusion.

John and Elizabeth Macarthur would have dressed much like this unknown couple in 1793.

This picture, attributed to Juan Ravenet, is included despite its poor
It shows a reception for the officers of the Spanish expedition that visited
house, from where the New South Wales Corps was to set off to
boy are John Macarthur

quality because it is the only image of street life in early Sydney.
in 1793. The site is the parade ground in front of Major Francis Grose's
depose Governor Bligh fifteen years later. It's possible the man and
and his son Edward.

Reverend Samuel Marsden, who in 1794 found Parramatta a scene of everything immoral and profane. *John Macarthur, its military commander, treated him* in a manner unbecoming a gentleman. *The two became bitter enemies.*

Governor John Hunter
Before his appointment he had never run anything more complicated than a ship. He took the job with a dream of comfort and satisfaction *and was to be bitterly disappointed.*

Governor Philip Gidley King and his family in 1799. King was perturbator *and* master worker of the puppets he has set in motion.

a good governor but no match for Macarthur, whom he called a
Macarthur outmanoeuvred King thanks to his superior patrons in Britain.

Lieutenant-Colonel William Paterson
When Macarthur shot Paterson, his commanding officer, in a duel at Parramatta in 1801, it was to change early Australian history.

Sir Joseph Banks
The 1808 rebellion had one of its roots in the enmity between Macarthur and the influential Banks, who nominated William Bligh for the position of governor. Bligh arrived in New South Wales determined to crush Macarthur.

Sydney in 1802. Note Bridge Street, the broad diagonal road in the

right-hand third of the picture, running up to Government House.

Governor William Bligh
As a mixture of old-style autocrat and modern bureaucrat, Bligh was both behind his times and ahead of them. While good at obeying orders and giving them, he had trouble dealing with his social equals.

Major George Johnston, leader of the rebellion of 1808.
Much maligned as the tool of Macarthur,
Johnston was actually a competent soldier driven by Bligh to take extraordinary measures for the good of his regiment.

Governor Lachlan Macquarie
After dealing with Bligh for a short time, he concluded:
It is impossible to place the smallest reliance on the fulfilment of any engagement he enters into. *Honouring one's word was the mark of a gentleman.*

thought otherwise. Indeed, he became obsessed with the issue of his status, so that Newton was to write to him in May 1795 that he had *laid too much stress upon your office as chaplain, and the right you had from thence to be considered and treated as a gentleman.* Johnson had been prepared to travel halfway round the world to live among the biggest collection of sinners it contained. But to expect him to be treated other than as a gentleman was asking too much.

At the same time as Johnson was complaining about Grose, the acting governor was assuring London how well the colony was going. The crops were flourishing, three large mills were at work, and he had built barracks for the troops. The amount of land cleared during his command was 2,962 acres. (By that month, April 1794, the total cleared since 1788 was 4,665 acres.) He mentioned this not, he said, because much credit was due to himself, but *because I wish to represent in the most favourable point of view the officers serving in the colony, to whose great exertions the promising appearance of it may be entirely attributed.* He noted, somewhat insouciantly, that he had not yet carried out Dundas' order to reduce the number of convicts given to each officer, as they had cleared so much land and grown produce *of much public utility.* Among the most enthusiastic were the Macarthurs, who in April 1794 received a second grant of 100 acres, adjacent to Elizabeth Farm. The next month, Elizabeth gave birth to *a very fine boy, to whom I have given his father's name John. He, with the other two, Edward and Elizabeth, are in perfect health, and promise fairly to become everything we could desire.*

In August 1794 Macarthur could tell his brother back in Plymouth that *the changes we have undergone since the departure of Governor Phillip are so great and extraordinary that to recite them all might create some suspicion of their truth.* Elizabeth Farm now covered 250 acres (another 50 acres had been bought from their owner). Henry Dundas, in a letter to the new governor, John Hunter, in England on 1 July said that the government could not afford to pay Macarthur for his work as inspector of public works, but that he could receive extra land or convicts instead. Strangely enough, there is no evidence that this in fact occurred. However, even the ability to ensure the convicts allocated to him were good workers would have been a great help. One hundred of the Macarthurs' acres were now under cultivation, with the excess produce (£400 worth in the past year) being sold to the commissariat. Most of their remaining acres had been cleared of timber, and they had three horses, two cows, 100 hogs and 130 goats, all, apart from the cow given them by Grose, purchased or bred. One man and six greyhounds worked in the bush to put wild duck and kangaroo meat on their table.

The colony around them was also growing and flourishing. A letter written by a soldier at this time described how, after disembarking at Parramatta from the Sydney boat, *On your ascending the wharf appears a row of huts on each side, and a spacious road to the distance of a mile; at the upper end, Governor Phillip erected his country seat. The garden that surrounds it is beautiful, abounding, in the season, with grapes, melons, pumpkins, and every other fruit and vegetable. The florist may also amuse himself … the botanist, I believe, may here find the most beautiful shrubs and evergreens that produce very fragrant flowers. The governor's garden at Parramatta is so situated by nature that, in my opinion, it is impossible for art to form so rural a scene.* The Spanish botanist Née had written about this *magnificent garden, in which are grown all the vegetables of Europe and various trees which serve it as a useful ornament. The pomegranate trees were not very large, but were well loaded with fruit; strawberries served as a border for the paths, and in these geraniums and white stock [also] grew luxuriantly.*

In another part I saw the vineyard.

The unknown soldier quoted above also noted that many boats had been built in the colony, *which supply us with plenty of fish, and the oysters are the largest I ever saw.* The kangaroos had *retired up the country* (presumably those closer in had all been eaten), there were plenty of chickens, pigs and goats, and spirits were now plentiful: *the gentlemen always purchase the cargoes, and this watery mixture is sold at 16s. per gallon.*

In the same month he wrote to his brother, Macarthur drew £756 in paymaster's bills to finance his investment in a trading voyage of the *Britannia* to the Cape. Macarthur was the second biggest investor. The biggest was John Palmer. Others this time included Captains William Paterson, Joseph Foveaux and George Johnston, Surgeon John Harris and Ensign John Piper. In between their farming and trading activities, the military officers continued to guard the colony. In October at Parramatta some Irish convicts stole a six-oar boat owned by Macarthur. Two days later it was found down at Botany Bay.

In December Grose took an important step when he made the first land grants on the Hawkesbury River, near present-day Windsor, then known as Green Hills. The Hawkesbury has several major tributaries, including the Nepean River, whose sources rise in mountainous country far to the south and west. The soil of the river flats around Green Hills was far richer than any yet discovered in the colony; a crop of potatoes could be planted and dug up in three months. Only later did the settlers realise that this rich loam was the product of occasional, devastating floods, which could appear unexpectedly on a fine day because it had rained the day before over the mountains. But the soil was so good the area quickly became a magnet for ex-convicts, attracted by both its remoteness from authority and the ease with which it could be farmed. This area was added to the responsibilities of Macarthur, who advised Grose on the land grants. James Ruse, whose luck had turned bad at Parramatta, led twenty-two farmers into the wilderness. The Hawkesbury was quickly to gain a somewhat lawless reputation, and to be the scene of many fatal encounters with the Aboriginal inhabitants. It was to provide a lot of produce as well.

On 17 December 1794 Grose, still suffering from the wounds he had received years before in North America, followed Phillip's lead and made a unilateral decision to return to England, even though John Hunter, Phillip's successor, had not yet arrived. He left a letter assuring him that *The colony is at this time in so flourishing a state, and the officer I leave to command every way so capable of the duty of it, that no evil consequences can possibly attend my going away*. He suggested that Hunter would find the settlement improved since he was last there, and added modestly that most of this had been Phillip's doing.

Grose was a generous leader, generous with others' land and money. He had done nothing to stem the trading activities of the officers, or to deny them plentiful free convict labour. In all these activities John Macarthur played a major part, and, thanks to the enthusiasm with which he embraced the opportunities on offer, was one of the major beneficiaries of Grose's rule. In two years Australia had acquired an economy, and the Macarthurs had jumped from being almost penniless to a position of modest affluence. Grose was replaced by his deputy, the 39-year-old Captain William Paterson, an amateur botanist who in 1789 had published a book about his collecting journeys in South Africa. He had commanded the troops on Norfolk Island from 1791 to 1793, in his spare time sending a flow of botanical and geological specimens back to his patron, Sir Joseph Banks. On his return to Sydney he tried unsuccessfully to find a way across the Blue Mountains, as many more were to do after him. As the new acting governor, he pursued the same policies as Grose. In broad terms the Grose–Paterson period can be seen as the interruption of the rigorous naval culture of the early governors by a more lenient, free enterprise military culture.

1795

In March there would have been much rejoicing among the officers of the New South Wales Corps, for the *Britannia* arrived with forty-one good horses from the Cape of Good Hope. This was the first decent lot of mounts to reach Australia. Gentlemen weren't supposed to walk, they rode, so the arrival of these steeds had deep symbolic, as well as practical value for their new owners. All that status; all that manure. (The scarcity of fertiliser was the biggest single problem in early Australian agriculture.) In May Macarthur was promoted to the rank of captain and the *Shah Hormuzear* returned with 131 head of cattle from India. Only one cow had been lost, which was highly unusual: it was the cost and risk of bringing beasts to Sydney that kept their populations down for so long. But the colony was still in desperate need of more large, heavy-hoofed animals to carry its gentlemen and ladies, pull its ploughs, enrich its soil and put meat on its tables.

Whereas the settlement of Sydney and Parramatta had occurred with relatively little armed conflict, partly because of the death of many Aborigines in the 1789 plague, the Hawkesbury saw frequent fighting between blacks and whites. Of the four hundred settlers on the rich river flats stretching thirty miles along the river in mid-1795, five had been killed and several wounded by Aborigines in search of corn. Paterson sent two subalterns and sixty privates – a large detachment – to the river to drive the Aborigines away. The soldiers killed about seven Aborigines on their first night in the area, and captured five more. These were sent to Paterson who tried to convince them that it was *not in their interest to do us injuries* before releasing them. He had acknowledged in his report to London that he regretted having to kill the natives, *particularly as I have no doubt of their having been cruelly treated by some of the first settlers who went out there.* But he was prepared to use violence when necessary to ensure the settlement's survival. He stationed a small group of soldiers on the Hawkesbury permanently. In 1796 he issued an order warning the settlers that anyone killing a native except in self-defence would be guilty of murder. Some twenty-six whites were killed and thirteen wounded over the next four years. It was not recorded how many blacks were killed.

The officers increased the trade in spirits, and it is interesting to read the view reached by a doctor who later examined the health of the early colony. Noting the extent of hunger and disease during Phillip's time, Sir James Watt speculated that *rum, by supplying the missing dietary calories, provided the energy which finally broke the vicious cycles of malnutrition and its sequellae [of disease] to enable the colony to survive at the price of a generation of addicts*. Some might have been addicts before they reached Sydney.

Macarthur remained inspector of public works. During this period, while private agriculture thrived, the public activities were run down. Most of the convicts left on the government farms were badly fed, so that they were forced to spend their spare time working for private employers to survive. The amount of government land being farmed shrank, many of the convict labourers had no housing provided by government, and the one barn at Toongabbie was blown down and not rebuilt. New settlers, whether ex-convicts or the few free immigrants, were sometimes given land that proved useless for cultivation.

Macarthur seems to have abused his position as inspector of public works to some degree. He interfered with the departments of other officials, such as the chief surgeon. He ordered the commissariat to issue clothing and supplies to certain people and restricted those from whom it was prepared to buy grain: the store would not even take grain from the government farm if the officers had produce to sell. It forced small farmers to sell or barter their produce to the officers at reduced prices. It's likely Commissary John Palmer was involved in this scam, which would have included the soldiers guarding the store and the storekeepers who worked in it. Macarthur also probably allocated convicts to his friends and to himself in greater numbers than to others. As the provision of convict labour and the purchase of grain were, apart from land grants, the key transactions in the economy of New South Wales, these actions were crucial to the enrichment of Macarthur and some of the other officers. In much of his activity he was acting with the approval of Grose and Paterson and there were considerable long-term public benefits in creating a vigorous economy in New South Wales so soon. The question that must have been in his mind as he waited for Hunter's arrival was how he would persuade the new governor to approve the extraordinary changes that had occurred in the years since Phillip's departure.

★

How much of Australian history has been influenced by its origins? As noted in the Introduction, I believe the answer is more than is often recognised, and here I would like to set out the case in a little more detail. I do so not with any hope of proving my claim, but simply to indicate why I think Macarthur was important. He matters because early Sydney matters, and he was a vital part of it.

Alexis de Tocqueville believed that nations, like people, are shaped by their early years. Just as *The entire man is, so to speak, to be seen in the cradle of the child,* so too *The growth of nations presents something analogous to this; they all bear some marks of their origin. The circumstances that accompanied their birth and contributed to their development affected the whole term of their being.* It's an interesting hypothesis. There are two ways in which it could be true, either as one event affecting another, or as our beliefs about the past affecting our behaviour in the present. Regarding the first way, the most important possible influence was social. The British class system began to die from the moment the boats landed at Botany Bay in 1788. It died far more quickly here than it did in Britain, and the story of that death is one of the great stories of Australia. A society was built in Sydney in the early decades which, being largely composed of convicts and ex-convicts and their children, could not sustain the British belief in a great chain of being, a genetic hierarchy of human worth. Britain transported convicts in the belief it could thereby rid itself of criminal types. But New South Wales proved this belief mistaken: the children of the convicts were generally honest citizens, while Britain continued to produce criminals in abundance. Free settlers in the colony were often less competent than emancipists. These things were not supposed to happen: the Australian experience was socially antipodean.

Thus, it can be argued, Australia was launched in its earliest days on the voyage to becoming a particularly egalitarian society. Other factors were to contribute later, but the journey began here, and it was by no means inevitable that this would happen. The historian Alan Frost has suggested that Sydney in the nineteenth century could have turned out as a class-ridden society in which all former convicts were ruthlessly kept down. That it did not owes something to the character of Arthur Phillip and the behaviour he introduced in those early years, including his insistence on the same rations for everyone in the colony and the promotion of some convicts to the position of supervisors. There would, of course, be plenty of attempts to suppress the emancipists over the coming years, and Aborigines were to be treated badly. Yet compared with the Old World, new ways were established by Phillip, and their product, Frost suggests, can still be seen in Australia today.

Such a view of early Sydney requires that we think less about convicts and their suffering and more about ex-convicts and their happiness. This is difficult for some: the popular notion of early Sydney is still dominated by the convict experience. However, most convicts were freed, and spent far more of their lives in Australia in that condition than as prisoners. Therefore it is worthwhile to pay it more attention. If we look at tolerance, today considered by many to be our greatest national virtue, we can see a direct line between the generally favourable treatment given to ex-convicts (land, labour, stores, and even, under Macquarie, dinner invitations), the emergence of democracy many decades later, the progressive welfare system of the early twentieth century, and the positive reception in the later twentieth century of immigrants from different cultures and races around the world.

Other indications of Australia's future can be seen in its beginnings. The strength of the labour movement in the twentieth century has an echo in early Sydney. In the following chapters we will see how convicts actually had considerable control over their labour, for some of which they were paid. In order to get the most out of them, the governors were forced to do many things that affected the development of New South Wales, including shortening the hours of unpaid work they performed and shutting down unproductive government farms. Convicts had rights: they could own property, give evidence in court, and complain to a magistrate if they were ill-treated. Ex-convicts often benefited from labour shortages and found better wages and conditions than workers back in Britain. Thus was the tradition of relatively non-subservient labour born here. E.P. Thompson has noted that the eighteenth century saw a move from paternalism to more objective relationships between employers and employees. New South Wales, where feudal traditions were non-existent and labourers independent, moved in this direction more quickly than Britain.

For a long time Australia was obsessed by the so-called convict stain. In recent decades this has been replaced by a sort of defiant pride in our origins as bad boys and girls, which in its own way is equally limiting to our idea of ourselves. It might be more useful to see the convict – particularly the ex-convict – experience as hugely beneficial because of its contribution to the democratic culture of this country.

Other national characteristics influenced by the early years include mateship and the relative independence of Australian women, which owe much to the preponderance of men in the early decades: many women could pick and choose the best men, while those many males left on the shelf were forced into each other's company. Then there is the national reliance on government. The British government ran the place for many years and poured large numbers of people and huge sums of money into it. The pleasures of dependence were learned early. The colony was run by an autocratic governor, often unpopular but usually a man of integrity with a concern for justice. There were no sources of money or power to seriously rival the government, which got into a habit of doing things for the population that was to persist for two hundred years. Because the government was fairly competent, and taxes were insignificant, there was little motivation to develop the ideals of independence and self-help so dear to Americans. That a majority of the population consisted of prisoners, ex-prisoners and public servants (including soldiers), all used to being told what to do, would have contributed to this. The importance of government was to continue largely undiminished for many years because the two greatest sources of wealth – land and free convict labour – were in its gift.

As noted in the Introduction, the strength of government in New South Wales occupied some of the psychic and economic space occupied by religion back in Britain. We have seen in the last two chapters some examples of the failure of religious leaders to make much of an impact here. Historian John Gascoigne suggests that the values of the Enlightenment were particularly ingrained in many of those involved in the settlement of New South Wales. And because there was no conservative tradition to react against these values, they were to assume a more dominant role in New South Wales than in Britain.

I think one other aspect of later Australia had its roots in the early years, although I suggest this more tentatively than I have the above links. Australian nationalism, when it developed many years later, was to be an unusual, non-classic nationalism, very much unlike, say, the French or German variety. It depended on racial and linguistic cohesiveness, yet in neither of these was it unique: we were not the only Anglo-Celts in the world who spoke English. Because of our obvious links to Britain and other parts (and ex-parts) of the empire, we have never been able to think of ourselves as a nation in the sense that, say, Japan or Poland can, as the main repository of a certain language and ethnic type.

I suspect this fact, rarely acknowledged, has created a great deal of angst over the years in those Australians who believe we should be more nationalistic than we are. This was manifested in the great disappointment with which some greeted the result of the 1999 republic referendum. But because we were part of the British Empire for so long, our nationalism, when it developed, was dualistic and diluted. We were used to being both Australian and part of something else – not necessarily part of an empire, but part of a race and a culture spread around the globe. In a way we were used to globalisation, on a small scale, before it became fashionable. Left-wing Australian nationalists have long decried this situation (or rather, its manifestations) but most Australians seem to like the dualism that exists. Where many intellectuals see a conflict between Australia and Britain, perhaps because they focus on political institutions and arrangements rather than cultural and ethnic ones, many other people see a complementarity.

While this was all extremely embryonic in the period covered by this book, it was nevertheless the case that British politics, and connections with powerful British patrons, were to be very important in the conflicts that John Macarthur became involved in. The sense of dualism, of living here but being very aware of there, was already present. The early history of Australia is very much part of the history of Britain, a point worth making before we plunge into the detail of the following chapters.

I acknowledge that all these theories depend on judgements regarding national character that can be, and in most cases have been, disputed. They have also been attacked by other, equally deterministic, ideas, such as Russel Ward's suggestion that the Australian character was mainly forged later in the outback. Perhaps we are on safer if less stimulating ground if we leave theories of objective influence behind and look only at how the stories we tell about our past might influence the way we feel about ourselves and our future. Memory is important whether it be true or false, and nations, like individuals, have memories. This is what history is. In this sense the above theories are important not if they're right (assuming this could be determined) but if they're part of popular consciousness. And most of them are not.

As far as most people are concerned, the big story about Australia's birth, particularly in recent decades, has involved little of the above. It's been the convicts, whether they were ignored or, in more recent times, embraced. The tension and frisson surrounding the convicts – the stain, the ignominy, the resentment, the growing pride – have to a large extent excluded any other story from the popular imagination. But, as I have argued, the convict experience was an aberration, even in the lives of most people who were themselves convicts, and actually tells us little about our past. We have given it more importance than it deserves. By seeing the convicts as victims and the free colonists as predominantly vicious and insignificant we trivialise our past, and cut ourselves off from much of the complexity and richness of our origins.

IV

Hunter

On 7 September 1795 John Hunter finally returned to Sydney. He'd been there at the beginning, when Governor Phillip had found Botany Bay unsatisfactory in January 1788 and gone north to discover Port Jackson. Later he was in command of the *Sirius* when it sank off Norfolk Island.

Hunter, whose father had been a captain in the merchant navy, was a Scot who had studied at the University of Aberdeen, although without graduating. He loved music and had painted some of the creatures and flowers of New South Wales, the results notable for both their charm and their scientific inaccuracy. His book about the settlement around Sydney and at Norfolk Island, published in London in 1793, had been translated into German and Swedish. While his career had been moderately successful, he was fifty-eight years old and had no experience of running anything more complicated than a ship. He got the job because of his patrons, and despite the fact that Phillip had recommended Philip Gidley King.

For reasons that are not known, it had taken Hunter a year to leave England after his appointment. In that time he received various instructions from Henry Dundas, at the Home Office, who was *concerned that great evils have arisen from the unrestrained importation of spirituous liquors into our said settlement from vessels touching there, whereby both the settlers and convicts have been induced to barter and exchange their live stock and other necessary articles for the said spirits to their particular loss and detriment.* This view of the colony probably came from the Reverend Johnson via William Wilberforce. In the King's name, Hunter was enjoined to crack down on this practice, or else to incur *our utmost displeasure*. He was also told to set aside land in each town for a church, with 400 acres to support the minister, and another 200 acres for the schoolmaster. The combination of a day job with farming was not to be restricted to the officers.

Before Hunter left England, Dundas was replaced by the Duke of Portland, who told the new governor that Grose had been wrong to persist with government support of the convicts he had so generously allocated to the officers. Grose's claim that this arrangement was beneficial for the colony was incorrect, Portland pointed out. The convicts should be fed by, and work for, either the government *or* their private master. But they shouldn't be fed by the former and work for the latter. (The duke would not be the last to notice the enthusiasm of Australian farmers to socialise costs and privatise income.) So the British authorities had strong views on the issues of spirits and convict labour, and Hunter was perfectly aware of these as his ship entered Port Jackson.

At the time of his arrival, Macarthur, Foveaux and Palmer were the three largest farmers in the colony. Macarthur seems to have gained Hunter's confidence as quickly as he had Grose's three years earlier. Hunter was largely ignorant of agriculture and finance, and relied on Macarthur in his early months in the colony – the captain came highly recommended by Paterson. They went over the country together, and Macarthur pointed to the public farm and said it was exhausted, *with scarcely an acre fit for cultivation.* He recommended that Hunter buy a large amount of grain from the private farmers in case of future need, even though the store was almost full.

Four days after landing, the new governor felt confident enough to write to Portland that the state of agriculture far exceeded what he had expected and did great credit to the regime of Grose and Paterson. He approved the emphasis that had been placed on private rather than public farming, for *If government were to continue to cultivate land sufficient for the maintenance of whatever number of convicts may be hereafter sent out … there would be an effectual stop to the exertions of industrious farmers for want of a market for their crops.* He retained Macarthur as inspector of public works, *a situation for which he seems extremely well qualified.* In late September the governor was a guest at Elizabeth Farm, staying overnight. Soon after he wrote again to Portland, describing how self-interested officers had been far more productive than government farms. Everything was so marvellous that he disobeyed his instructions and allowed the officers to retain the extra ten convicts – still clothed and fed from the government store – they'd been given by Grose. So government continued to provide the officers with free labour and then pay them for the produce; in effect, paying twice for its grain and meat. Even if Hunter did have other reasons for maintaining the status quo, one might have expected him to at least try to justify this financial black hole in his letters to Portland. But he had no taste for conflict. He had arrived expecting a cushy post, possibly because of his experience of the relative smoothness with which things had been run by Phillip. He later admitted, apparently without irony, that when he accepted the job he had had a *dream of comfort and satisfaction.*

Presently Hunter began to wonder about Macarthur's representation of the state of the colony. On the first Sunday after his arrival, he attended a service in the Reverend Johnson's church, where he heard a sermon which *exposed the last government, their extortion, their despotism, their debauchery and ruin of the colony, driving it almost to famine by the sale of liquors at 1,200 per cent profit.* Johnson then *congratulated the colony at the abolition of a military government, and the restoration of a civil one, and of the laws.*

The governor certainly noted the fact that he could find scarcely twenty convicts to perform public works, even if the full implications of this took a while to sink home. He pondered the fact that 150 settlers had been given farms under Grose and Paterson with no title beyond a piece of paper on which the commanding officer had scrawled that so-and-so had his permission to settle. Some of these settlers were convicts whose terms had not even expired.

One thing Hunter did do early was to return to civilian hands those magistrate's functions that Grose had transferred to the military. The three magistrates were William Balmain and Richard Johnson in Sydney and Samuel Marsden in Parramatta. Their duties extended beyond the bench, to matters such as checking the accuracy of weights and measures and supervising road works. Hunter also appointed Captain George Johnston as his aide-de-camp.

Of course there were other things going on in the colony

apart from its political and social birth pangs. Land was being cleared, kangaroos killed, crops grown, babies were being born, and everyone had to come to terms with the new country. Not all were as positive as Elizabeth Macarthur. The convict artist Thomas Watling observed that *the landscape painter may in vain seek here for that beauty which arises from happy-opposed off-scapes*. He liked contrasts, but in New South Wales found only that *everything approaches or recedes by a very striking and singular gradation of proximity*. The wildlife suffered from the same degree of similarity, he felt. All those grey marsupials. On the other hand, the Aborigines were interesting, more intelligent than many thought yet possessing noteworthy quantities of *irascibility, ferocity, cunning, treachery, revenge, filth and immodesty*. Watling wondered why Hunter fed some Aborigines while there were white people who were *frequently denied the common necessaries of life!*

In December Hunter issued a government and general order to the populace, explaining that the government was paying too much for its grain and asking the farmers *to consider what reduction of the price of wheat and Indian corn they are willing at present to submit to*. He explained that if the price was not reduced, the government might have to grow its own or even ask the farmers to provide their convicts with bread. Had this been followed up with effective action, it would have been admirable, but as with most of his efforts at reform, it petered out.

Less than a month after Hunter's arrival, a convict searching for game to the south-west of Sydney was taken by an Aborigine to an area of native grassland where he found a large herd of cattle. This was very exciting. The convict told his master, who told Hunter, who in November set off with a party towards the area now known as Camden. There they found some sixty-one wild cattle. They tried to shoot a calf but were charged by a bull, which dropped only after taking six shots. These beasts were the survivors of those that had been lost back in 1788, and their descendants. Hunter called the area the Cowpastures. This was a marvellous discovery, and by 1800 the herd had increased to some three hundred. Apparently the Aborigines, who later were to spear cattle around the continent, generally ignored this growing herd. Even stranger to relate, the Europeans had no way of catching the beasts, despite their desperate need for cattle.

In November Hunter found a convict, George Hughes, who could use the small printing press that had been brought out with the First Fleet but had lain idle ever since. Australia's first printery was set up in a room in Government House, used at first only to issue the governor's orders to the population. Also this year the Macarthurs had another child, Mary.

1796

By this stage there were 4,016 colonists in New South Wales, and the society was becoming more complex. A trickle of free settlers was arriving, and many of the convicts had served their sentences and were now free men, whether farmers (all so-called emancipists were entitled to land grants), tradesmen, labourers or criminals. Due to the war with France, relatively few new convicts were to arrive until 1814, so the proportion of free men rose steadily throughout this period, creating a burgeoning new class with interests that the governors had to try to come to terms with.

The society was developing culturally as well as commercially. On 16 January Judge Advocate David Collins attended a theatrical performance of *The Revenge*, put on at a playhouse set up by *some of the more decent class of prisoners, male and female*. He found their performance *far above contempt*; they had been warned that *the slightest impropriety would be noticed, and a repetition punished by the banishment of their company to the other settlements.*

One of the problems facing Hunter was the deteriorating condition of the Tank Stream, Sydney's source of water. It was a poor supply and tended almost to dry up during droughts. In late 1791 Phillip had ordered holes, or 'tanks', to be cut into the rock to hold water that would otherwise run into the harbour. Three tanks were cut where Spring Street now meets Pitt Street. One held almost 8,000 gallons. As the settlement increased, the stream became polluted, and Hunter put a fence on each side of the stream and banned people from keeping pigs nearby, or using dirty vessels to collect water. But despite the punishments that were threatened, primitive industries were set up by the stream and people attempted to dam or even dig wells next to it, and the contamination continued.

The soldiers of the New South Corps were intimately involved in the life of Sydney. Those who were not single lived outside the barracks, often engaged in business on the side. Many lived with convict or ex-convict women; some of the locally recruited soldiers were emancipists themselves. In early February one of these, a member of Macarthur's company who when a convict had suffered under a supervisor named John Baughan, discovered his old boss at work in a building nearby. Putting down his musket, he engaged in conversation with an acquaintance, its subject matter considerably abusive of Baughan. The ex-supervisor overheard this, as he was intended to, but then the plan of revenge backfired. Baughan sneaked out the back of the building and took the musket, which he carried off to the man's sergeant. The soldier was relieved of his post for the not inconsiderable offence of *being without his arms*.

At 9.30 the next morning, a large group of angry off-duty soldiers approached Baughan's house. He grabbed a gun but the soldiers surrounded the fence, broke it down, and overpowered him, forcing him onto the ground where one axe-wielding private threatened to kill him if he moved. A labourer named George Wilson, also armed, tried to help Baughan, but his musket was wrenched from his grasp and used to smash the house's window, after which it was broken over the fireplace. The military mob, with Baughan's terrified wife standing by, proceeded to destroy all the couple's furniture and then the very house itself. Once their work was done, they went off cheering.

Governor Hunter issued an order describing the incident as *disgraceful to the character of the British soldier* and expressing his disappointment that none of the soldiers had offered up the names of *the ringleaders of so mutinous a conduct*. He discussed the matter with Paterson and said he would address the soldiers himself, but later in the day changed his mind, deciding this would be a condescension on his part that they did not deserve. He told Paterson the soldiers' behaviour was *the most violent and outrageous that was ever heard of by any British regiment whatever* and assured the commander the culprits would be brought to justice.

Baughan identified the ringleaders to Judge Advocate Collins. Many free colonists were outraged by the event. Assistant Surgeon William Balmain, a civil magistrate, met with Baughan, who expressed his fear of what would happen to himself if the soldiers were prosecuted; indeed, his wife told Balmain she believed he would be killed. Balmain told them not to worry about this, and added that Baughan would be compounding the felony by refusing to proceed. Indeed, it would be an offence, and Balmain might charge him with it.

Hearing of this advice, John Macarthur entered the affair. The officers of the New South Wales Corps were enraged by what they saw as Balmain's interference in their area of authority. Their dominance of the colony for almost three years under Grose and Paterson had not broadened their appreciation of the principles and practice of the rule of law. Macarthur wrote to Balmain asking if what he had been told about the meeting with Baughan was correct. Balmain replied that it was. On 10 February all the officers wrote to Balmain expressing their indignation at *his shamefully malevolent interference in the affairs of the corps*. The letter was in Macarthur's handwriting. Balmain then wrote to Paterson, complaining that the officers in general had been drawn into the dispute by the base insinuations of some particular person, who *delights in strife and who would meet from [me] the chastisement he merits if [I] could fix on him*. When Paterson did not reply (he passed the letter on to his officers), Balmain wrote to Macarthur, again accusing him of creating the dissension and calling him *a base rascal and an atrocious liar and a villain*. This was effectively a challenge to a duel. The letter was delivered by John Palmer, the commissary, who had agreed to be Balmain's second.

Was Macarthur to accept the challenge, thereby acknowledging that Balmain was on the same social level as himself and his brother officers? The gentlemen of the colony discussed the matter carefully. The next day they again wrote to the surgeon, informing him that *Mr Balmain's letter to Captain Macarthur had been read to them, and that they were unanimously of opinion that no other notice could be taken of it by him but that which he did take*. In other words, nothing. These were still the days when doctors were not as effective, and therefore not as respected, as they later became. A mere Mister had presumed to challenge a Captain, and the largest group of gentlemen in the colony had rejected the presumption. They noted that Balmain's letter proved *him ignorant of the language of a gentleman, as his language have determined him incapable of sentiments of honour or integrity*. They generously conceded that should Balmain choose to *resent in a proper manner the opinion of the corps towards him*, they would appoint one of their number to fight him, and if that soldier went down they would send forth *another and another*. It was an insulting rejection of Balmain's assumption to the same social status as themselves, reflecting the way the concept of honour was not fixed but in any particular circumstance could mean whatever powerful people wanted it to mean.

Balmain wrote back on 11 February seeking to separate Macarthur from his brother officers and assuring them of his intention to *maintain to his last breath the character of a gentleman in defiance of every unmanly mode of detraction that Mr Macarthur is capable of using*. And that was it – there was nothing more he could do.

Having seen off Balmain, Macarthur now approached Hunter with an offer. The soldiers involved were very sorry, he said, and would behave better in future. If the whole matter could be dropped, they would pay Baughan for the damage. The men had already been given a stern lecture on the affair by their superiors. This request was backed up by a personal petition from Baughan (one can guess at the pressure that had been brought to bear on him) to the governor, asking that the warrants for the arrest of the four ringleaders be withdrawn. Hunter did exactly as Macarthur had asked and, in the proclamation he issued seeking to explain this extraordinary decision to the anxious people of Sydney, he even warned the convicts to stop being insolent to the soldiers. It was a terrific victory for Macarthur and must have boosted his reputation within the corps, and the wider colony.

Balmain was, not surprisingly, upset by this outcome, and by his treatment from the corps. On 7 March he told Hunter *how very difficult it must at all times prove for any man to do the duty of a civil magistrate when, by the proper discharge of that duty, he is liable to the reproaches and insults of a body of officers*. Balmain observed that *Mr Macarthur's propensity to turbulence and litigation has ever been conspicuous in this colony*, and *he has not himself scrupled to avow his inclination to be contentious, and to undervalue the power of the civil authority*. In this episode the civil authority – the governor and the law – was defeated by the need to defend the honour of the corps.

However, the incident seems to have brought Hunter's growing suspicion of Macarthur to a head. On 25 February Macarthur resigned from his position of inspector of public works, claiming that the governor had lost confidence in him and had allowed convicts to be removed from his control by third parties without informing him of this. Hunter said that what had really happened was that Macarthur had got into the habit of pretty much running things as he liked, and resented the governor's decision to restore some of this power to the civilian officers. Clearly, the trust Hunter had shown in Macarthur in his early days had at least been modified. And the power and influence of the civilian officers was rising accordingly, to Macarthur's chagrin. He had been told that Richard Atkins, at the governor's dinner table, had asserted *that the most glaring partiality had been exercised in receiving maize at the stores*. Macarthur denied it and told Hunter how offended he was that such things should be allowed to be said without rebuke in Government House. The governor replied that he had never heard anyone refer to this at his table, and expressed his concern that *there should be a single character in this colony who felt more delight in having an opportunity of creating distrust and discord than of promoting confidence and harmony*. On the same day as he wrote this, he appointed *Richard Atkins Esq.* as the new inspector of public works. Atkins' brother, Admiral Sir George Bowyer, was a good friend of Admiral Lord Howe, a former First Lord of the admiralty and Hunter's patron.

As so often with Macarthur's disputes, it is impossible to

know exactly what was going on in this case. Perhaps he had become aware that Hunter was about to sack him and resigned as a pre-emptive strike, using the story that Hunter had allowed Atkins to make allegations against him at table as a justification. As we shall see, it was Macarthur's technique in a dispute where he felt vulnerable to hunt around for some new issue where he might portray himself as wronged, and then use this to cloud the original cause of the argument. He also took great pleasure in raising the stakes in a quarrel, forcing his opponent to go further than originally intended or to withdraw.

What are we to make of Atkins' allegation, assuming it was in fact made? Given the wealth Macarthur accrued during this time, it's quite possible he was engaged in rorting the commissariat, although there's no direct evidence of this. During this decade and the next one, other people made similar allegations against the officers in general. They claimed that small farmers were prevented from selling their produce to the store and thereby were forced, to acquire necessary goods, to exchange their produce for items whose value was wildly over inflated. For instance and using hypothetical values, let us say that a shipment of hinges had just been received by the commissary from England, to be sold to anyone who wanted them for two shillings a pair. The officers would inspect the whole shipment first and buy up anything of value, including all the hinges. Assume you're a small farmer who has just harvested four shillings' worth of maize. You take it to the commissariat, hoping to get in return a pair of hinges and two shillings in credit. But the commissary refuses to accept or pay for your maize, saying he has enough. If you want the hinges, you now have to buy them from an officer trader (or his representative) – who has marked them up by 100 per cent, so they cost you all your maize. Such behaviour seems to have been frequent from the early 1790s, and made money for many of the military and civilian officers, and those who worked for them. However, the evidence is circumstantial.

In March Hunter wrote to Portland about the prevalence of the illegal distilling of alcohol, presumably a form of vodka, using wheat. *I find it has been in practice here some time, and appears not to have been carried on with much secrecy. Many of the settlers and others have fallen upon a method of erecting stills and distilling a most noxious and unwholesome spirit, which not only serves to destroy the health of those who use it, but it also consumes a quantity of grain which would otherwise come to market.* These letters of Hunter's – and there were to be many of them – are curious because he seems to be describing an appalling state of affairs as an outsider, ignoring his own responsibility for its continuing existence. There is a peculiar air of helplessness in much of what he wrote.

On 17 July 1796 Richard Atkins, in his capacity of Parramatta magistrate, wrote to Macarthur informing him that on two occasions soldiers had been caught stealing turnips from the garden of Government House there. He said he did not want the man caught most recently punished, but hoped it would not happen again. Macarthur responded in a letter addressed to *Mr* Richard Atkins, rather than the more respectful title of Richard Atkins *Esq*. He insisted on being told the culprit's name. Atkins refused to provide this, as he had told the soldier he would not report him. *I must observe*, he noted sententiously, *if rigid justice is the order of the day, the Lord have mercy on us all.* He concluded: *It is further, sir, necessary for me to inform you that any letters directed to Mr Richard Atkins will be returned unopened.*

Macarthur immediately took up his pen and wrote to Hunter, claiming that in refusing to provide the thief's name

Atkins had treated Macarthur *with insufferable insult.* Macarthur, whose thirst for justice must have increased since the Baughan affair, asked His Excellency to make Atkins surrender the soldier's name. Hunter directed Atkins to do so, but told Macarthur that his use of *Mr* in addressing Atkins was wrong as it appeared to have been intended *to mortify him as a gentleman, or to lessen him as a magistrate.*

Macarthur replied with an extraordinary letter that marks the beginning of his reputation for excessively aggressive behaviour. He struck back fiercely, expressing his astonishment at the very suggestion that he might have intended to offend Atkins through *the use of a trifling appellation.* Such a thing, Macarthur suggested, could hardly affect the reputation of *a man so deeply plunged in infamy.* How, he asked, could such a man *be imagined to possess one feeling of a gentleman when the enormities he is committing daily are considered*? It was degrading to His Majesty's service that an officer such as himself should be so insulted by *a man so vile ... a public cheater, living in the most boundless dissipation, without any visible means of maintaining it than by imposture on unwary strangers whose business leads them to this settlement.* It was a letter to make a governor's heart sink. And there was more. Hunter, naturally, would want evidence. Macarthur said he had it. *I will prove that in his public and official capacity drunkenness and indecency are almost inseparable from him; and that no longer since than the 19th instant he was exposing himself at an early hour of the morning in the public streets in the most disgracing state of intoxication.* Macarthur felt His Excellency needed to know *how exemplary a character the distribution of public justice is entrusted to.* (See Appendix for copy of these letters.)

Hunter asked for details, and these were provided; they included claims that Atkins had twice given someone a bill of exchange to be drawn on London agents who declined to pay them, had robbed a settler in the public highway, and had asked for a letter he'd written to a military officer to be destroyed lest it harm his (Atkins') reputation. Macarthur closed this list of horrors by referring to Atkins as *a man so publicly branded with the commission of the vilest frauds, and the practice of the lowest vices*.

Why did he decide to insult Atkins in this manner? The circumstances of his resignation from the position of inspector of public works, and Atkins' elevation to it, suggest one or more causes. Hunter later suggested another reason to Portland. Atkins was about to replace David Collins, who was returning to England, as judge advocate, and Hunter believed Macarthur had sought to blacken his reputation so he could not do this. The implication was that Macarthur would have preferred the position to be kept in the hands of the army. Hunter saw this as part of a broader attempt by some of the military officers to undermine the growing role of civilians in the colony's administration, Macarthur's attack on Balmain being an earlier example.

Judge Advocate Collins passed Macarthur's list of charges to Atkins, who eventually replied, denying all of them. But first he wrote directly to Macarthur, opening by saying that the repugnance he felt in doing so had been overcome by *the gratification that naturally results to a man of honour in delineating even by paper the deformity of an opposite*. It is a fine letter, if a little elaborate by today's standards, a masterpiece of abuse in which (unlike in Macarthur's own attempts at the genre) the self-righteousness does not restrain the venom. Atkins told Macarthur he was a man *the baseness of whose heart even imagination, however warm, can hardly portray*. Still, he very sportingly gave it a go.

Macarthur had said in a letter to Collins that his *charges* against Atkins might be more properly called *assertions*. Atkins described this as *screening yourself by a jingle of words from that manly perseverance which should mark the character of a man professing as you do. The quibble was only worthy of a dastardly coward like yourself. Your original meanness*, he continued, slicing closer to the social bone, *and despicable littleness pervades your every action. It shows the cloven hoof. Return to your original nothing; we know what you have been, and what you now are; and believe me an honest and industrious staymaker [a reference to Macarthur's father] is a more honourable and more useful member of society than such a man as I hold you to be.*

Moving on, Atkins then accused Macarthur of being *the incendiary – who has been the promoter of all the feuds and animosities between individuals in this colony*. He thought Macarthur's allegations against him presumptuous, given that the captain *four years ago, was only a lieutenant, penniless but by his pay, and is now reputed worth £8,000. ... On this subject, viper, you bite a file; the day of retribution will come, and believe me it is not far off, when you will be dragged forward by the strong arm of justice to public view as a monster of society*. Finally he wrote that Macarthur had *been called liar and scoundrel, and, if I am not misinformed, received a blow.*

You'd think such a letter would have provoked an immediate request for satisfaction, and Atkins told Hunter this was his intention. But Macarthur did not respond in this manner. We know he was no coward, but as his response to Balmain's challenge shows, he had decided to ration the number of insults he took seriously. Perhaps he realised it would do him no good to shoot such a well-connected man as the new judge advocate. But he was worried by the postscript to Atkins' letter, in which the magistrate informed him that his list of charges, *as well as some other well-authenticated facts*, were being sent *home for the perusal of those whom it may concern.* Given Atkins' contacts in England, this might be serious.

Macarthur's response was to raise the stakes even further in what was becoming a potentially dangerous situation for him. On 13 September he decided, presumptuously, to bypass Hunter and write to both his own commanding officer, Lieutenant-Colonel Grose, and to the Duke of Portland, about the dispute with Atkins. But he went even further in the letter to Portland, by complaining about Hunter's administration of the colony. Whether he did this from sheer irascibility, or because he knew Hunter would take Atkins' side and wanted to damage him first, the risks were considerable.

Before writing to Portland, Macarthur laid the ground. On 18 August he put in his first really big ambit claim for more government assistance. He wrote to Hunter and reminded him that he had once suggested that officers and others with convict farm labourers ought to provide the bread for their men. This would save the government the expense of feeding privately employed convicts. Macarthur now repeated the suggestion, acknowledging the *very liberal indulgences* he and others had received from the government, and offered to provide his own convicts with bread from now on. He also offered to take on another hundred convicts on the same terms, promising that he would be able to support them with bread and meat in eighteen months or sooner.

Hunter replied that this was the first time anyone had made this offer. He accepted it as it pertained to Macarthur's existing convicts, but declined the invitation to provide him with any more. Macarthur responded hoping that his example might *stimulate others to exert themselves in promoting the general interests of the settlement*, and asking for some clothes for his servants and 10,000 nails from the store, as he needed to erect some buildings to accommodate the astonishing increase in his livestock. He got the clothes but not the nails.

Now Macarthur wrote to Portland, describing his generous offer to take a hundred convicts off the government bread supply. (That the governor did not actually have a hundred convicts to spare him was ignored.) He told Portland, as he had told Hunter, that it was crazy for the government to both support convict labourers on private farms and buy the product of their labour from their masters. He claimed that he knew more about farming in New South Wales than anyone else, and that his proposal would save the government a lot of money. He then proceeded really to put the knife in, saying he had made this proposal to Hunter when he first arrived in the colony, but *it is painful for me to inform your Grace that this resolution has never yet been put in practice*. He further had to point out that in the colony *the interest of government is utterly disregarded, its money idly and wantonly squandered, whilst vice and profligacy are openly countenanced*. He could prove *that many who are fed at the expense of his majesty are permitted to dispose of their time as they think proper themselves; that others are allowed for the service of convict prostitute women; and that the labour of one-half of the people is directed to purposes which can never be of the smallest utility or advantage to the colony*. He went on to list various other problems, informed Portland of the *melancholy truth … that vice of every description is openly encouraged*, and attached to his letter a plan for encouraging the increase of livestock.

It was another extraordinary letter. Other colonists, such as the Reverend Samuel Marsden, would have concurred with much of its contents, while pointing out that Macarthur, as one of the most powerful men in the colony for the past four years, might have had something to do with the situation so evocatively described. Macarthur gave the sealed letter to an outraged Hunter to forward to London, and the infuriated governor wrote a covering note to Portland, pointing out that *were every restless, speculating, troublesome, or dissatisfied individual in this or in any distant colony encouraged to consider himself of suf-*

ficient importance to take the liberty of corresponding with his majesty's ministers, well, *it would soon occasion such a variety of opinions as could only serve to embarrass the judgement of government, and would generally be found to be directed more to the private interest of the several schemers than that of the public.* Hunter noted that he did not possess a violent or a peevish disposition, but he considered Macarthur's conduct to be *an impertinent, indirect, and highly censurable interference* in his own position and duties.

So these letters went off to London, together with a bundle from Judge Advocate David Collins relating to the Atkins affair, which included some character references organised by Atkins' ally Samuel Marsden. The vicar assured the Home Office that he had never seen Atkins drunk – at least, not when performing his duties as a magistrate. Marsden himself bore a grudge against Macarthur, who, he later told Hunter, had sought privately *to assassinate my character* and publicly to destroy his reputation with *violent and shameful attacks*.

What happened to Macarthur's letters is perhaps the strangest part of this strange affair. Portland received them and then, almost a year later, sent them back to Hunter, asking him to respond to the charges they contained. He agreed that Macarthur should not have written to him directly, but said nothing of a reprimand. And, worst of all, he ended his letter by saying that he had so high an opinion of Hunter's *penetration and judgement as to leave no doubt upon my mind of your availing yourself of every suggestion contained in Captain Macarthur's letter, which appears to you to tend to the advantage of the colony, and the diminution of the public expense.* The correspondence was to drag on, but essentially Macarthur had got away with criticism of the governor that could have been treated as serious insubordination. Portland was starting to be very concerned about Hunter's competence and profligacy with the government's funds. Macarthur's letter had reinforced these concerns, and in not reprimanding its author, Portland was sending Hunter a polite but clear message. But Hunter didn't get this – what continued to concern him most in the affair was the simple fact that Macarthur had gone over his head. For a man who had spent his whole life on ships, where a commander's orders were obeyed without question, such behaviour was hardly conceivable. He became obsessed by it, and was to refer to Macarthur and his action many times in his subsequent letters to Portland. Yet he never seems even to have begun to think of what he might do to respond effectively to the letter. Philip King was later to say Hunter told him he'd challenged Macarthur to a duel but Macarthur declined. We know no other details, but this was the most likely time for the challenge to have occurred.

In this year Edward Macarthur, aged seven, was placed on board the *Marquis Cornwallis* and sent alone to England to be educated. It would have been an extraordinarily wrenching experience for Elizabeth and possibly John, neither of whom was to see him for many years. That he was sent, and at such an early age, indicates the importance of a good education in a young gentleman's life, and also, perhaps, his parents' desire to remove the growing boy from the pernicious influence of New South Wales.

In August 1796 Hunter wrote to Portland begging that the New South Wales Corps be recalled. He suggested it had been a mistake to recruit from *the dregs and refuse of our native country* rather than from the *very best and most orderly*, and described the corps soldiers movingly as *a set of the worst, the most atrocious characters that ever disgraced human nature … often superior in every species of infamy to the most expert in wickedness amongst the criminals.* We know the men in the corps' ranks were no worse on average than those in the rest of the army, so this fulmination must have been inspired by the decline in his relationship with the officers. He predicted correctly that the government would decline to follow his advice due to its cost. Hunter was a man who loved to be right. On a more positive note, it was now that he wrote the despatch urging the government not to send any more incompetent free settlers to fill government positions (*the millwright and miller ... are sad impositions*) but to allow those already in New South Wales to fill the jobs, even if they were convicts.

About now Elizabeth Macarthur received a letter from her old friend Bridget Kingdon, apologising for writing so infrequently but pointing out that only one of the many letters she had sent had ever reached Elizabeth. The British newspapers contained many favourable reports of *N.S.W.* (as she called it), and she wanted to visit but would not, due to *the dread and danger of the voyage*. She tried to give an account of the latest fashions in women's clothes, but like a good daughter of the vicarage admitted that it was not really her subject. She had observed, though, that there was no longer such a thing as a waist, and stays could now be dispensed with. *The petticoats are up to the shoulders and below the feet. Hats are still worn small, but indeed dress as you will you are not thought particular unless you have a long waist, and it is very seldom that we see two people whose garments are made the same. Tis quite fashionable to write on coloured paper, but you my dear friend will I hope excuse my using plain white, indeed this place will afford no other.*

In September 1796 William Paterson left for England and Joseph Foveaux replaced him as acting head of the corps. Macarthur assumed the position of paymaster, previously held by Foveaux, the extra responsibility more than doubling his salary. In November Hunter again wrote to Portland complaining about the New South Wales Corps in general and John Macarthur in particular. Possibly thinking of the ancient Greek practice of ostracism, he wondered if there should be legal means of *removing from the colony such people as are found fond of disturbing the peace and order of the settlement, and of private society*. As New South Wales had been settled as a place to which such people might be sent, rather than be sent from, it was an impossible dream. In the same month David Collins left Sydney along with John Palmer, who was returning to England on leave. Hunter sent a letter with them to Under-Secretary John King, saying the two men would inform him of the *mischiefs and embarrassment, vain, restless, and designing people may occasion in a country peopled with such characters as we have here*. He assured King that Collins and Palmer would tell him *of whom I complain*. It was Macarthur, of course, and again Hunter urged that *government would direct that such disturbers of the peace and happiness of a distant country like this might be sent out of it*. Then he said, *You will be told in England that a fortune of eight or more thousand pounds have been made here in five years. A question naturally arises out of that information – How is it possible?* He attacked the prevalence of trade and the fact that, on his arrival, he'd found that the military could take whatever they wanted from the stores – *whole casks of nails at a time* – until he put a stop to the practice.

Hunter presented himself, here as in many of his letters to London, as a great reformer. Yet under his command public farming actually stopped for a while in the colony, and he refused to follow Macarthur's sensible suggestion of making settlers support their convicts. One thing he was certain about, though, was windmills, something of a private enthusiasm, and he told Portland at the end of 1796 that he was *erecting upon the high ground over Sydney a strong substantial and well-built windmill with a stone tower, which will last for 200 years, and we are preparing materials for another such at Parramatta.* These would be used to grind grain for flour.

With David Collins' departure, the colony's gentlemen were now split into two camps, with the civilians clustered around Hunter in Government House on the hill to the Tank Stream's east, and the military officers in their houses and barracks on the western side. It might seem strange that Atkins had been appointed to replace Collins, given his various faults and lack of legal knowledge, but high birth and contacts counted – in 1800, much closer to London, John Toler was appointed chief justice of Ireland, despite his *scanty knowledge of law, his gross callousness, and his buffoonery*.

1797

In June a piece of paper containing a denunciation of Hunter was found in the streets of Sydney. Anonymous pamphlets like this, known as *pipes*, were to be a feature of the colony for some years. They were common in Britain, and much feared by those in power, possibly because they provided a voice to the voiceless. The identity of their authors was often sought with the offers of rewards. This one accused Hunter of sanctioning his servants' involvement in the spirit trade, and so upset him that he asked publicly for anyone having evidence of the claim to come forward. He also offered a reward for the identification of the author of the *infamous, scandalous* letter, but was unsuccessful on both counts.

Hunter was not happy as governor. In all his despatches to London there is evidence of an inexhaustible capacity to brood and worry, without any corresponding desire to address the causes of these concerns. His letters to the Home Secretary read more like diary extracts, ones void of self-knowledge, than the reports of an effective administrator. Their great theme is his dissatisfaction with the amount of time and energy the military officers put into trade rather than their army duties. Perhaps the best summation of this came in a letter of June 1797, when he opined that *this spirit for trade, which I must ever consider in the manner it is carried on here to be highly disgraceful to men who hold in their hand a commission signed by His Majesty, has been carried so far that it has now reached all the inferior appointments, so that it has absorbed all their time and attention, and the public duty of their respective offices are entirely neglected.*

David Collins' lively history of the colony's early years, published some years later, covered some of the period after he left it. He recorded that at this time *dogs had increased to such an extent as to occasion their becoming the object of a public order, restricting the number kept by each person to no more than were absolutely necessary for the protection of his house and premises. Much mischief had been done by them among the hogs, sheep, goats, and fowls of individuals.* By this time Sydney had become a bit like a scruffy English town. There were three schools, and just before they broke for Christmas, Collins reported, *the governor was gratified with the sight of 102 clean and decently-dressed children, who came with their several masters and mistresses, and in form paid their respects to his excellency, who examined the progress of the elder scholars in writing, specimens of which he kept for the purpose of comparing with those they should present to him on the following Christmas.*

On Christmas Eve two settlers on some land around Homebush Bay had a drinking competition using raw spirits. *Their abilities, however, were not equal to their boasting,* Collins noted, *for one of them died upon the spot, and the life of the other was fast ebbing* when he was found.

1798

By this time the Macarthurs had between 400 and 500 acres, including 120 acres of wheat, and fifty cattle, twelve horses and about one thousand sheep, and lots of pigs, goats, geese and ducks. In this year, according to Elizabeth Macarthur, the Macarthurs used a plough for the first time in Australia. Ploughs had actually been brought out with the First Fleet, but not used until now, possibly due to the hardness of the unmanured soil and the scarcity of large animals to pull them. Convicts with hoes were much easier to obtain, even if often they just scratched the ground.

Because of the war in Europe there continued to be few convicts from England, so that the incoming Irish comprised an increasing proportion of the colony's population. This created considerable anxiety among the free people. David Collins wrote that the Hibernians were *so turbulent and refractory* that only the most severe treatment could extract any work from them at all. *In addition to their natural vicious propensities, they conceived an opinion that there was a colony of white people, which had been discovered in this country, situated to the south-west of the settlement, from which it was distant between three and four hundred miles, and in which they were assured of finding all the comforts of life, without the necessity of labouring for them.* Equipping themselves with the picture of a compass drawn on paper to help guide their way, a party of sixteen Irish were about to set out for this heaven on earth when stopped by the soldiers. Seven of them received two hundred lashes. When even this didn't discourage the Irish, Hunter sent a few of them off to the south-west with a party of guides and soldiers, to prove to them that no settlements were to be found in that direction. After a few days the exhausted convicts begged to be allowed to return to civilisation. Collins described some other act of Irish stupidity and cruelty, and concluded with the rhetorical question: *Could it be imagined that at this day there was existing in a polished civilised kingdom a race of beings (for they do not deserve the appellation of men) so extremely ignorant, and so little humanised as these were, compared with whom the naked sav-*

ages of the mountains were an enlightened people?

In July 1798 all the military and most of the other officers signed an interesting contract with each other. They agreed that *on the arrival of any ship or ships in any of the ports within the limits of this government having goods, wares, or merchandises on board for sale, two officers be chosen and nominated from amongst ourselves who shall, on the part of the whole, be authorised and empowered to treat with the captain, commander, or master thereof for the purchase of such goods, wares, and merchandises.* They further agreed not to buy anything from the ship outside this arrangement, which included not buying anything from a ship which had refused to deal with their agents, or had refused to accept their price. Anyone who broke the agreement would be fined £1,000. *And we further bind ourselves and engage for ever afterwards to avoid the company of any individual, and to consider him an infamous character, who shall be convicted of a breach of these articles.* A week later Hunter issued a general order assuring the population that anyone in the colony could buy *their proportion* of any such purchase from the cartel, *the officers having undertaken the trouble of officiating as agents for the general benefit of the whole economy.*

The effect of these two documents, assuming they were obeyed, would have been that the officers' monopoly operated for the benefit of all the inhabitants, by preventing ships' captains from charging extortionate prices. We don't know if Hunter's order was obeyed, but later contemporary criticisms of the officers' monopoly suggest it might not have been. The early governors had almost no staff, so it was difficult for them to know what was going on, let alone enforce unpopular change.

For months Hunter had brooded over Macarthur's *scandalous and bombastical letter to his Grace by a man who has employed the whole of his time in this country in sowing discord and enriching himself by means truly disgraceful.* When he finally received a copy of the letter from Portland, he immediately issued an order restricting the number of government-fed convicts per officer to two, as he had been instructed to do years before. And then, in July 1798, Hunter sent Portland his reply to Macarthur's letter. He claimed that Macarthur had been boasting in the colony that his letter to Portland would lead to Hunter's removal. Naturally, Hunter rejected all of Macarthur's claims against himself, and accused Macarthur of having lied to him when he first came to the colony and telling him that the government farm was unfit for cultivation. As a result, it was not sown with seed, and extra grain had to be purchased from the free farmers, including the officers. Yet later Hunter had it sown and it produced a fine crop.

Seeking further evidence against Macarthur, Hunter asked Surgeon Balmain to write an account of Macarthur's attempt to intimidate him over the Baughan affair, but Balmain replied that he had subsequently reached an agreement with Macarthur to forget the whole thing, and felt obliged by his *honour as a gentleman* to keep his word. Hunter had more success in drawing from the Reverend Johnson and his colleague Marsden bitter criticisms of the New South Wales Corps and their effect on the colony. He closed his letter by assuring Portland that *the sacred character of our Saviour, were he to appear in this colony in its present state, would not be secure from the dark attack of those whose private views he might oppose in favour of the public interest.*

In September, Elizabeth wrote to Bridget Kingdon, noting that the colony had been without mail for a whole year because a British ship had been captured off the coast of Brazil. *We firmly believed that a revolution or some national calamity had befallen Great Britain*, she wrote, *and we should be left altogether to ourselves.* New South Wales, she suggested, was *the only part of the globe where quiet is to be expected.* She raved about the good soil and climate of the colony, and the numerous advantages it offered anyone holding a government appointment. The only drawback was the difficulty of educating children: *My desire is that they should see a little more of the world, and better learn to appreciate this retirement.* Despite her obvious enthusiasm for the colony, she acknowledged that its society was *narrow* and noted that her children *all speak of going home to England with rapture. ... They have early imbibed an idea that England is the seat of happiness and delight; that it contains all that can be gratifying to their senses.* One wonders where they got this from. Would it have been from the mother who went on to say that one of the advantages of a British education would be to *undeceive* them of such notions? The emotional complexities of immigration.

It was spring and Elizabeth observed with delight *the most beautiful variegated landscape. Almonds, apricots, pear and apple trees are in full bloom. The native shrubs are also in full flower and the whole country gives a grateful perfume.* There was now a very good carriage road between Parramatta and Sydney, and another from Parramatta to the Hawkesbury, which Elizabeth had visited on horseback. She had passed through country the greater part of which *is like an English park, and the trees give it the appearance of a wilderness or shrubbery, commonly attached to the habitations of people of fortune, filled with a variety of native plants, placed in a wild irregular manner.* She spent a day on the river, going as far as Richmond Hill, which she climbed to look at the view. *I could have spent more time there,* she wrote, *but we were not without apprehensions of being interrupted by the natives, as about that time they were very troublesome, and had killed many white people on the banks of the river.*

Despite their relative success on the land, the Macarthurs still rationed their milk and had not yet even killed a sheep, so desperate were they to breed more livestock, the shortage of which continued to cripple the colony. Elizabeth said they had thirty or forty people working for them, of whom only two were fed *at the expense of the Crown, although there are persons who contrive to get twenty or more, which the governor does not or will not notice.* The Macarthurs had a piggery and fed their workers on pork, as well as buying various necessities of life for them from visiting ships, *for shops there are none.* The free workers *demanded an enormous price, seldom less than 4s or 5s a day.* She explained the cartel system of buying goods from visiting ships, which arrangement *prevents monopoly, and the impositions that would be otherwise practised by masters of ships.*

She continued: *I have had the misfortune to lose a sweet boy of 11 months old, who died very suddenly by an illness occasioned by teething. [He had been named James.] The other three, Elizabeth, John, and Mary are well. I have lately been made very happy by learning the safe arrival of Edward in England. We often remember and talk over in the evening the hospitalities which we have both received in Bridgerule Vicarage.* She asked to be remembered to a military man of their acquaintance, who would, she knew, be happy to hear that John (or *Mr Macarthur* as she called him in her letters) had found in New South Wales *a field for advantageous exertion.*

Elizabeth then moved on to the subject of Bridget's single state, asking *Are you difficult to please? Or has the war left you so few bachelors from amongst whom to choose?* She offered some advice: *Abate a few of your scruples, and marry. I offer in myself an instance that it is not always, with all our wise foreseeings, those marriages which promise most or least happiness prove in their result such as our friends may predict. Few of mine, I am certain, when I married thought that either of us had taken a prudent step. But look at them now. At this time I can truly say no two people on earth can be happier than we are. In Mr Macarthur's society I experience the tenderest affections of a husband, who is instructive and cheerful as a companion. He is an indulgent father, beloved as a master, and universally respected for the integrity of his character.*

His trading activities were continuing, too, although we don't know what profit he made from them. Between 1792 and 1798 the officers of the New South Wales Corps drew bills to the total value of £36,844 and invested it. The money was spread around most of the officers, and Macarthur was among the more enthusiastic.

Presently Bridget wrote back to Elizabeth regarding matrimony, asking what *unfortunate females* such as herself should do *who have not an offer from a person they can approve*? Warming to the subject, she asked what Elizabeth would have her do, *not surely be so eccentric as to reverse the matter, and make an offer (if you would) I have not courage, nor vanity enough to pursue the scheme, unless indeed I had a vast deal of the ready, now so much looked after, and indeed so absolutely necessary*. Resigning herself to spinsterhood, she congratulated Elizabeth on her happier fate, and told her how charming dear little Edward had been when he spent a week at the vicarage.

1799

In February Portland wrote a furious letter to Hunter explicitly ordering him to arrest any officer found engaged in trade and return him to England to face court martial. The next month Portland wrote to the commander-in-chief, who wrote to Lieutenant-Colonel Paterson, then in England, regarding the trading abuses being practised by the officers of the New South Wales Corps, in which the profit received *is often at the rate of one hundred shillings for one*. This wild exaggeration might also have come from the colony's vicars via William Wilberforce. Paterson was informed of the order for Hunter to arrest those guilty, and ordered back to the colony *in the hope that your presence will materially contribute in some measure to restore the credit which has hitherto attached to the character of a British officer, but which has in this instance been sullied*. This appears to be the first time the army's commander-in-chief had become involved in the situation in New South Wales. Taken with Portland's direct order for Hunter to send home the trading officers, one might have expected that the glory days of Macarthur and his colleagues were now numbered. But Hunter simply ignored the order.

In May the governor, now sixty-one, wrote to his friend Sir Samuel Bentham, brother of Jeremy. He noted that before he left England he had seen a model of the proposed ideal prison, the panopticon. *I might almost venture to say that had I been shut up in one of those cabins or cells and kept hard at work for the same length of time I have spent here that my comforts could not have been less than they have been, but my peace of mind would certainly have proved considerably greater*. He said that he had been intentionally embarrassed in the performance of his duties, and had realised (at last!) that the military officers had been disappointed that the governorship had passed to *another naval officer*. The salary looked good on paper but was entirely absorbed by the expenses of the job. He noted bravely that he would persevere with his miseries until his health and strength began to decline, which should be soon, at which point he would ask His Majesty's permission to resign.

A group of British missionaries who had been driven out of Otaheite (modern Tahiti) by its inhabitants now fetched up in Sydney. They were appalled by what they found. *Here*, they wrote home, *we have to contend with the depravity and corruptions of the human heart, heightened and confirmed in all its vicious habits by long and repeated indulgences of inbred corruption, each one following the bent of his own corrupt mind, and countenancing his neighbour in the pursuit of sensual gratifications*. The vices that were *fashionable and familiar* included *adultery, fornication, theft, drunkenness, extortion, violence and uncleanness of every kind.* While it is tempting to dismiss this as the hypersensitive reaction of missionaries suddenly returned to civilisation, it's a useful reminder that New South Wales was full of criminals and was, compared with Britain, a moral cesspit. But the British authorities were not overly concerned – that, after all, was the point of its existence, to be the home of the unredeemable criminal class of Great Britain.

One of the missionaries was murdered in July. Brother Clode had befriended a soldier named Jones and his wife, to whom he lent thirty shillings. He asked for it back and the couple killed him. Various officials, including the governor, came to view the corpse, and a trail of blood led helpfully to the Jones' residence, where was found an *axe which had been newly washed and not quite dry, with the signs of blood, brains, and some human hair upon it.* The couple and an accomplice were executed, and the two men were left hanging in chains while the woman was dissected.

The officers in New South Wales and their families would have been constantly aware that their position was envied by the large numbers of criminals among whom they lived. The French Revolution had found some resonance in the hearts of at least some of the Irish and the British lower classes. In mid-1799 a convict ship named the *Lady Shore* left England carrying more recruits for the New South Wales Corps, and according to the purser at the start of the voyage, the soldiers on board were *the most disagreeable, mutinous set of villains that ever entered into a ship.* Some were Irish and French deserters, who had been brought on board at Gravesend in irons but then released and given guns. When the ship's captain complained about this, General Francis Grose, the former ruler of New South Wales, came on board for a few minutes and announced that the men might keep their weapons. He said he would come back the next day but, once on shore, he left for London and was not seen again. A mutiny had already been proposed and exposed, and the soldiers had told the man who'd informed that they intended to throw him overboard during the journey. When the ship was off the east coast of South America, some Irish and French soldiers rose up and bayoneted the captain and shot the chief officer. The mutineers put twenty-nine people, including some women and children, into the ship's longboat and cast them adrift. Fortunately, they reached Rio Grande, on the coast of Brazil, without further loss of life. This incident was extreme, but mutiny was not unusual: the soldiers at Norfolk Island had mutinied in a minor way; in January 1796 seven convicts had been killed in a failed attempt to seize the *Marquis Cornwallis* at sea; and in Sydney in August 1798 the Court of Vice-Admiralty conducted a trial relating to an attempted mutiny on the *Barwell*. And all the while the proportion of the feared Irish in the colony was increasing.

Finally, the Duke of Portland had had enough. On 5 November 1799 he wrote to Hunter sacking him. He ran through the many issues raised by the governor in his extensive and repetitive correspondence over the past four years: the trading, the convict labour, the prevalence of spirits, the budget blow-out. Hunter appeared to be fully aware of the problems in each case, but utterly incapable of resolving them. It had taken him a long time to reduce the number of convicts allocated to the officers. And despite his constant fulminations against the trade in spirits, and his claim that it was debauching the lower orders, he had never issued a public order banning it, even though Portland had directed him to do so. Possibly Hunter shared the view of the Duke of Wellington that *Nobody in the army ever reads a regulation as if it were to be a guide for his conduct – or in any other manner than an amusing novel*, and felt orders would be a waste of time. But his superiors in Whitehall believed differently. Portland expressed his disapprobation of *the manner in which the government of the settlement has been administered by you in so many respects*, and told him he was to be replaced by Philip Gidley King. It would be wrong to assume that Macarthur's letter to Portland had led to the sacking; Macarthur himself was obviously implicated in many of the problems identified by the duke. But his letter and Hunter's response could only have increased Portland's conviction that, as a governor of New South Wales, Hunter was out of his depth. He could neither obey orders nor enforce them.

Of course it took a long time for this news to reach Hunter, but things did not improve in the meantime. Ten days after Portland wrote the above letter, the governor was putting pen to paper about the anonymous pipe, which had come to obsess him the way Macarthur's letter had. Unfortunately there was some truth in its contents. He acknowledged that his own *principal servant* had been caught mixing with criminals, indeed organising robberies of Government House. After Hunter confronted him, the man shot himself in the head.

Some of the officers continued to trade vigorously, employing their convict mistresses and servants to sell spirits and other goods over the counter, or door-to-door as *strolling peddlers*. Lieutenant Anthony Fenn Kemp, a company paymaster, built a shop at what is now the intersection of King and George Streets, and encouraged his soldiers to take their pay in goods rather than money. When a soldier turned up he'd be asked *What will you have? I have a very good tobacco at 10 shillings a pound, tea for 20 shillings a pound, and prints at 8 shillings a yard.* If the soldier insisted on money, Kemp would call him a *damned saucy mutinous rogue* and threaten him with flogging.

A court case of 1799 shows how Macarthur and the other officers operated to suppress those who dared oppose their power. Isaac Nichols was an overseer of government convicts who refused to allocate enough of them to the officers. H.V. Evatt argues in his book *The Rum Rebellion* that Nichols was framed on a charge of receiving stolen tobacco and brought before a court, which consisted of six officers and the judge advocate, Richard Dore. (Richard Atkins had been replaced in May 1798 by the newly arrived Dore, who actually possessed some legal training but was not a strong man.) The most outrageous hearsay evidence from some of the most disreputable men in the colony was accepted by the court. Macarthur and one of his servants gave evidence against Nichols, who was convicted by a vote of four to three. One of the three was a naval lieutenant then in Sydney, by the name of Matthew Flinders. After the trial, Flinders suggested it was a fit-up and Hunter wrote to England begging for an *upright and independent judge*. He would not be the last governor to make this request, or to have it ignored.

The court had sentenced Nichols to fourteen years' transportation to Norfolk Island, which Hunter suspended pending an appeal to England. Evatt claimed that Macarthur was later involved in trying to solicit evidence against Nichols from a convicted criminal. This appears to be a rare but interesting glimpse of the lengths to which Macarthur could go to defend his interests.

In 1799 at the Hawkesbury a soldier killed a black woman and her child, and Lieutenant Thomas Hobby, in charge of the troops there, heard that large numbers of blacks were to come down from the Blue Mountains for revenge. The situation was tense and an ex-marine named Goodall was speared while at work alone on his farm along the Parramatta–Hawkesbury road. An Aborigine thought to have been involved was captured and sent to the governor in Sydney. Hunter released him but was reported to have said that Hobby should have *punished him on the spot where he was taken*. Soon afterwards, three natives visited the settlement at the Hawkesbury, one of them carrying a musket that was identified by some of the settlers as having belonged to a farmer recently killed by blacks. The three men were taken by the settlers and their hands tied behind their backs. That night, a group of five vigilantes took the Aboriginal men outside. One escaped, but the other two were killed.

The bodies were found by a constable and Hobby arrested the five men, who were put on trial. The court found all the men guilty but it divided on what their punishment should be. As Hunter appears to have been in favour of punishing the men, the court referred the matter over his head to the Secretary of State, much to the governor's chagrin. The accused were released, and eventually pardoned.

1800

At the beginning of this year the colony, apart from Norfolk Island, had 5,100 people, of whom 2,546 were at Sydney, 1,466 at Parramatta and Toongabbie, and 1,088 at the Hawkesbury. The large number at Sydney suggests the colony was diversifying from its agricultural base, although Sydney was swollen by about 450 soldiers and their families. Of the total, 3,665 were still fed from the public store.

In this year Elizabeth Macarthur gave birth to her last son, William, a brother for another son who had been born two years before and named James in memory of the child who'd died in 1797. John Macarthur handed over the paymaster's job to William Cox, a newly arrived officer, who proceeded to build up a deficit of nearly £8,000 due to his trading and other private activities with the regiment's money.

Hunter prepared a list of the sort of work done by the colonists in the past year, which gives us some idea of the activities in the colony and the range of services provided for the officers. The carpenters repaired the inside of Lieutenant-Colonel Paterson's house and fitted out the interior of Major Foveaux's place; the shipwrights built a boat for Surgeon Balmain while the plasterers plastered and whitewashed his house; the tailors made 517 pairs of trousers; the bricklayers built a stable for Judge Advocate Dore and a new barracks 100 feet long and 18 feet wide; the blacksmiths made 508 hoes; five full-time men collected shells, cleaned them and burned them for lime; and the sexton tolled the bell and dug the graves. There were also two cutlers (one employed full-time by the surgeon), a ropemaker, a saddler, a bellman to give public notices, two basket-makers, a bookbinder, a watchmaker, a printer and an executioner.

In 1800, as the trees fell and the Macarthurs' wealth slowly increased, they had almost 300 acres of cleared and fenced land and another 1,000 acres of 'grass farm'. Clearing land was a lengthy and laborious process. Trees had to be cut down and burned, and stumps preferably pulled out. If left in, they occupied up to one-tenth of the land area. In the early days it had been hoped the stumps would eventually rot, but they did not. Macarthur estimated that it took one settler with one convict servant a month to clear an acre and plant it with wheat.

In January the *Swallow* arrived with good news about the progress of the war back home. The captain set up store *to gratify the inhabitants particularly the females*, David Collins noted, and sold *many elegant articles of dress from Bond Street [in London], and other fashionable repositories of the metropolis*. The women of the colony would have been able to buy the new, high-waisted dresses copied from the fashions of revolutionary France. Later in the month a Dutch ship called the *Thynne* arrived. It had been chartered by the military and civilian officers through their Calcutta agent. Macarthur, along with William Balmain and the acting commissary James Williamson, wrote to inform Hunter of this fact. They had the honour to tell him that yet more spirits were about to enter the colony, along with other goods. They pointed out the splendid benefits this offered, as any spirits brought in by other ships now could be purchased by *the lower description of inhabitants*, which would prevent a repetition of *the unfounded charges that have been made against the officers of monopolising*. Hunter dashed off a letter to Portland expressing his by now habitual bewilderment, saying he couldn't stop the spirits from being unloaded even if he wanted to, as no one would obey his orders. Also, another ship, the *Minerva*, was in port and Hunter had received a petition from eighteen people outside the officers' ring, including Simeon Lord, requesting permission to buy spirits from it with which to pay their labourers. He couldn't deny the emancipists without denying the officers, so he gave in to everyone. The *Minerva* purchases were the beginning of the end for the monopolists, as their total control over imports began to be diluted. It marked an important stage in the rise to prosperity of certain emancipist traders.

The numbers of Irish among the colonists continued to increase as shiploads of troublesome croppies arrived. (The rebels of 1798 cut their hair short – *cropped* it – to express sympathy with the French Revolution.) Many had been sent without records of the lengths of their sentences, and when inquiries were made it was discovered that the British authorities in Ireland did not even have records of their trials – if, indeed, they had been tried at all. Some had been transported illegally. Hunter noted that many had been *either bred up in genteel life or to professions unaccustomed to hard labour. We can scarcely divest ourselves of the common feelings of humanity so far as to send a physician, a formerly respected sheriff of a county, a Roman Catholic priest, or a Protestant clergyman and family to the grubbing-hoe or timber carriage*. Even some Irish could be gentlemen, apparently.

On 16 April Philip Gidley King arrived in Sydney and Hunter learned for the first time that he was being replaced. He reacted with his usual mixture of resentment and indecision. There was a perfectly good ship leaving the next week, but Hunter informed King that this would be too much of a rush, and he would have to wait for the next suitable outgoing vessel, in August. In the meantime, he refused to house King or transfer authority to him.

King, of course, had spent some half-dozen years in the colony previously, most of them on Norfolk Island. He wrote to Sir Joseph Banks in May that *my arrival and remaining here gives general satisfaction, but I believe many will change their tone when their nefarious proceedings are arrested. Vice, dissipation, and a strange relaxation seems to pervade all descriptions*. He noted that Lieutenant-Colonel Paterson, on his return from Britain at the end of the previous year, had issued an order to his officers to stop dealing with spirits and had then arrested Captain George Johnston, Hunter's aide-de-camp, for paying his sergeant with alcohol, while ignoring this behaviour from everyone else. *Strange to tell*, King added, Hunter refused to allow Johnston to be tried by a court martial, and he was sent home for justice.

It seems that Johnston had used his position to prosper. Apart from his transactions with spirits, King implied that he stole cattle from the government herd, and was the wealthiest man of the colony.

Like Hunter, King had been sent to New South Wales with orders to clamp down on the officers' abuses. Macarthur must have brooded long on this, for in July he made an important decision: he and his family would sell up and leave the colony. It seems he considered its golden age was over. He offered his land and everything on it to the government for £4,000. This was a lot. In England a good cottage could be built for £50. Three years earlier, William Wordsworth had taken out a lease on an old mansion with a deer park in Somerset for £30 a year. King felt he had to seek permission from London for the purchase, but he was very keen on it and accepted Macarthur's valuation. Macarthur had fifty cattle, ten horses and six hundred sheep, most of them superior beasts. His farm was so close to the growing town of Parramatta that King thought it should never have been granted away, and his grass farm was the best in the colony. King believed the proposed purchase would be a bargain for the government. Possibly the opportunity of getting rid of Macarthur also appealed to him. The Macarthurs and he began the long wait for a response from London.

Meanwhile, King was gathering information from various people around the colony, passing some of it on to Hunter, who became increasingly agitated as the time for his departure grew close. He was distressed that King appeared to believe many of the bad things he was told about the state of the place, and by King's determination, already stated publicly, to send home any officer who continued to trade in spirits. As King had no more – or less – power to enforce a ban than Hunter, any success he might have would reflect badly on his predecessor. So Hunter moped, and took offence, and wrote angry letters to King. It was a long and painful goodbye, finally over when Hunter sailed away on 28 September.

A few days before he left, he recorded the major public works completed during his time as governor. Examples included: two strong stone windmills in Sydney; two log granaries, one for wheat and one for maize, at the Hawkesbury; a strong wooden bridge over the Duck River between Sydney and Parramatta, *capable of bearing the weight of several heavy laden carriages at one time*; a church at Parramatta; a clock tower in Sydney; and the *large and elegant* Government House in Parramatta. There were also some works in progress, including a large water-wheel at Parramatta, and all the public buildings in the colony, decaying when he arrived, had been coated with lime and whitewashed.

By the time Hunter left the colony, its thirty-four officer-farmers had obtained, by grant or purchase, 14,584 acres of land, of which 1,528 acres were planted with crops. They also had 250 convicts and 6,295 head of stock. The most important economic transaction for much of early Australian history was the granting of land, but only for certain people. Had the young Macarthurs come here at any other time, they would almost certainly have not fared so well. But they had been lucky, and as the nineteenth century opened they were looking forward to returning to Britain with a modest fortune. All that was needed was for the Duke of Portland to agree that the Crown should buy their farm.

Hunter returned to Britain with his reputation damaged, and rightly so. But it was the fault of those who had selected him for the job: there had never been any reason to think he would do it well. Had King been sent out as governor five years earlier, as Phillip had recommended, the generosity of the Grose–Paterson years might have been wound back more easily. It was, of course, to the Macarthurs' considerable benefit that this had not occurred. Whatever happened from now on, they had traded their way up, and made the all-important jump from penniless gentlefolk to prosperous landowners.

V

King

King was a good choice as governor: experienced, reasonably tough, and fair. He came from Launceston in Cornwall and his father, like Macarthur's, was a draper, so he too was a first-generation gentleman. King was forty-two years old and suffered from what was described, perhaps inaccurately, as gout, which made him bedridden and irascible on occasions. The symptoms were *an almost fixed compression of the lungs and breast, with a difficulty of breathing and a constant pain in the stomach.* His wife Anna and their children accompanied him to New South Wales this time. (Previously he had had two sons by Anne Inett, a convict.) He was aware of the resistance he would face as he cracked down on the officers' trading, and made them pay for the food of most of their convicts, and he tried to build political support back home by warning his patron, Sir Joseph Banks, and his superiors of this in advance.

A curious feature of King's early governorship was that he had not been given a commission or any written instructions. He surmounted this potential difficulty by simply making them up, issuing commands to subordinates in which he quoted from fictitious orders from His Majesty. In a series of orders he proceeded to attack the basis of officers' wealth. The trade in spirits was outlawed, although the officers could still import sufficient to pay their own workers. But any officer trading in spirits from then on would be court-martialled. We don't know how much King's reforms bit into the profits of officers like Macarthur, but in any case the great trading days of the New South Wales Corps were drawing to a close, as emancipists like Simeon Lord expanded their businesses and free settlers of substance arrived. In June 1800, Robert Campbell, representative of Campbell, Clark & Co. of Calcutta, applied for permission to import goods into the colony. Campbell had been keen to break into the officers' monopoly since he first visited Sydney in 1798, and was also interested in the whales and seals that were common along the coast. These were by this time attracting the attention of whalers from America and Britain, and were to offer more profit than trade in the coming years.

Campbell provided damaging competition for the officers, as he came with excellent foreign contacts and considerable capital. He quickly ate into their dominance of the economy by offering customers better terms, in particular longer credit and lower interest rates. The officers did not like Campbell. At the same time, King was hurting them with economies, cutting the inflow of money from the British government. Most convicts now had to be supported by their masters. The commissariat cut back hugely on its purchases of farm produce, and on its gifts of tools and other supplies to the free settlers.

Although Australian governments would continue to provide lucky individuals with public benefits over the next two hundred years, never again would an entire class profit to the extent the several dozen officers of early Sydney had done. So well had some fared that, in King's words, they had *made sufficient fortunes to be indifferent about what happens to them* (i.e. to his reforms).

In June, King instructed Captain Joseph Foveaux, who was about to leave Sydney to run Norfolk Island, to establish a government store there to supply the settlers with essentials, which the Crown would mark up 30 per cent on perishable and 15 per cent on non-perishable goods. A store was subsequently opened in Sydney along the same lines. These measures, while not wiping out free enterprise, cut further into the profits available to the colony's traders by providing competition.

King's reforms were extensive, at least in theory. Rules were introduced for dealing with the behaviour of all ships in the harbour. Copper coins were brought from Britain and put into circulation. He issued an order insisting on respect for the Sabbath, and *a general attendance on Divine Service … All sentinels and watchmen are to confine every person who may be strolling about the towns of Sydney and Parramatta during the hours of Divine Service.* It's doubtful that he had any more success than Hunter in improving the settlement's morals. A few months later the Reverend John Youl, visiting Sydney, reported back to headquarters at the London Missionary Society that *no other spot on the face of the habitable globe contains more witnesses of the awful depravity of human nature. Sin, like a mighty torrent, overspread the land.* He thought the place ripe for a mission, and believed the Aborigines *would willingly give up their children to be educated after the manner of the English.*

King was concerned about the colony's white children. He

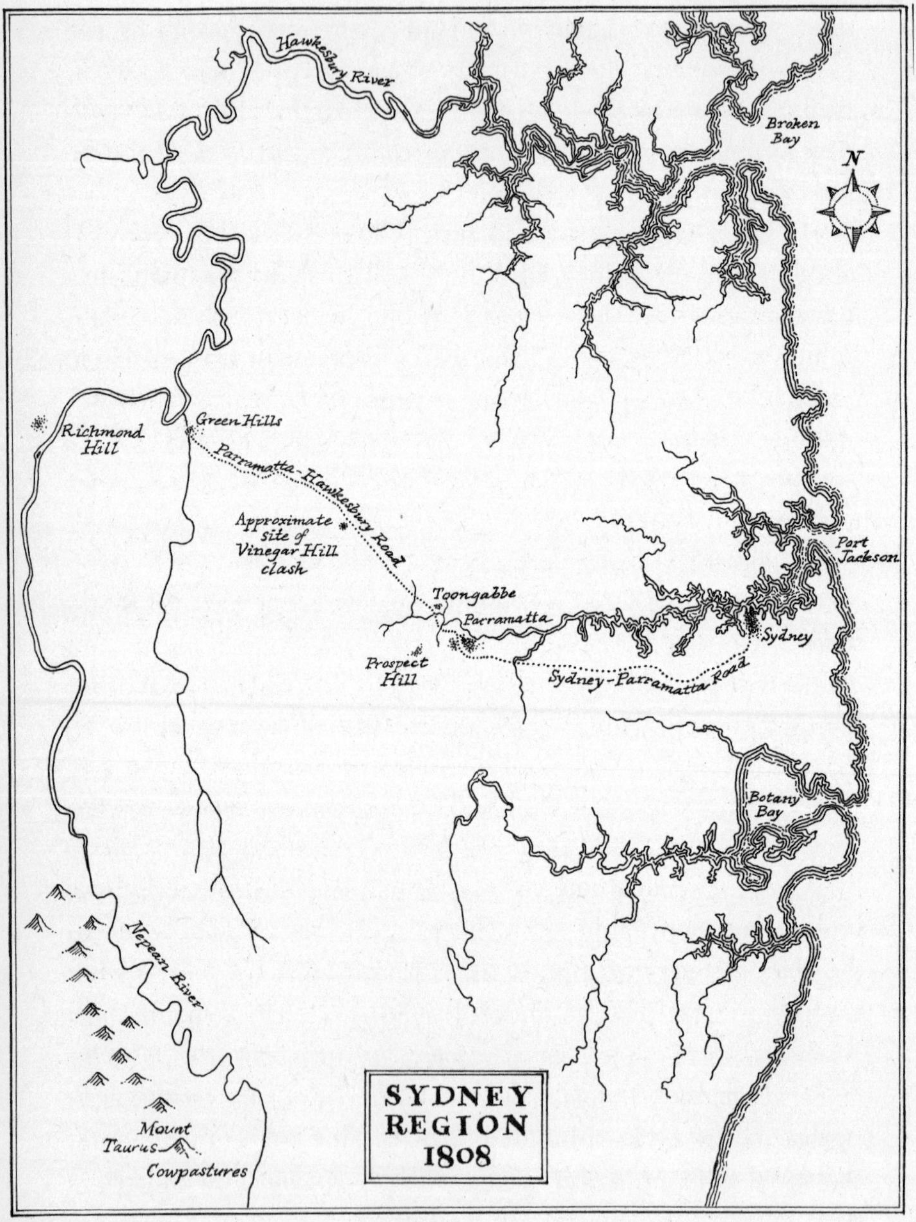
Hawkesbury River
Broken Bay
N
Richmond Hill
Green Hills
Parramatta-Hawkesbury Road
Approximate site of Vinegar Hill clash
Toongabbe
Parramatta
Prospect Hill
Port Jackson
Sydney
Sydney-Parramatta Road
Botany Bay
Nepean River
Mount Taurus
Cowpastures
SYDNEY REGION 1808

noted that soon after his arrival *I had frequent opportunities of observing the numerous children of both sexes going about the streets in a most neglected manner. This observation was confirmed by the many distressing relations made to me of the early abuses the female part suffered.* In May 1800, although not yet governor, King had offered to buy the house of Lieutenant William Kent, reputedly the best in Sydney, for the impressive sum of £1,539. Kent was returning to England and his home became Sydney's orphanage, housing some seventy children. Building work began later that year on a larger orphanage at Parramatta. King had established one on Norfolk Island some years earlier, when he was the lieutenant-governor there. He ordered Foveaux, when he reached Norfolk Island, to use the islanders' taxes and any fines imposed on them to maintain the orphanage and open schools. *I cannot too much recommend to your most serious attention a regard being had to the education and protection of the youth within your command*, he wrote, adding that *the future welfare of these colonies* depended on education. In Sydney he likewise used taxes and fines to fund the orphanages, and formed a committee to administer them. Its members were the Reverend Richard Johnson, William Balmain, the Reverend Samuel Marsden, John Harris, Mrs King and Mrs Paterson. The children slept two to a bed, each bed five feet by three. It was originally intended that the orphanages house children whose parents were dead or absent and those whom the committee judged incapable of caring for their offspring. However, when Portland heard of this he told King that he should *oblige all those to maintain their own families who have the means of so doing*. Thus began Australia's long debate over state welfare. In 1800 the number of the colony's inhabitants on the mainland who were not on the store and were self-supporting passed 50 per cent for the first time.

In the first two years of King's time as governor, almost three times as many convicts arrived as had come in the five years of Hunter's administration. A large proportion of them were Irish political prisoners, captured after the unsuccessful revolt of 1798, which had culminated in a battle in Wexford at Vinegar Hill. Over a thousand Irish arrived between 1800 and 1804. Many were far more resentful than the other convicts. The first major challenge for King occurred in September when he received reports that an uprising was being planned. A group of officers led by Judge Advocate Dore and Lieutenant-Colonel Paterson interrogated the Irish priest Father James Harold, who had already told William Balmain of the plot. He refused to name anyone (possibly his information had been acquired during confession) but said the plan was to attack the soldiers at the outposts, seize their arms and march on Sydney. The convicts hoped to be joined by discontented small settlers on the Hawkesbury and unhappy sailors from a ship in the harbour. Harold said he had talked the convicts out of it, was confident the uprising would not proceed, and was sorry he had ever mentioned it to Balmain. Again he refused to name names, so the committee gaoled him *for prevaricating and withholding the truth.*

Other convicts were questioned and claimed the plan was to put King to death, and that it was hoped some of the soldiers would join the uprising. Then *they intended to live upon the farms of the settlers until they heard from France, whither the insurgents meant to dispatch a ship.* The convicts further said that the rebellion was due to occur soon because the rebels *were well provided with plenty of pikes.*

Pikes were a central element in this and the many subsequent uprising scares, probably more because they might provide objective evidence of rebellious intent than because of their efficacy as weapons (the word seems to have meant a spear). The convicts claimed that *they had pikes made at Parramatta*, and *that some pikes had been concealed at Kelly's farm.* The plot's dimensions ballooned with every fresh witness: it involved taking Parramatta and marching on Sydney; it was being organised by an unnamed soldier; the plotters expected the small settlers to join because *they were so crabbed at their wives being struck off the store.* Blacksmiths were interrogated – those pikes had to be made somewhere – and said they knew nothing. One man's evidence was *evasive, equivocal, and jesuitical.* Another made claims based on what he'd been told by *an elderly man with black hair at Parramatta, whose name he says he knows not.*

Most of the evidence was hearsay, but finally the committee found a man who claimed to have joined the plot, who informed them of the complicated method by which the rebels might recognise each other: *By placing the forefinger of the left hand in the palm of the right, and closing the same; that if this signal was understood, it would be acknowledged by the party challenged clasping two forefingers of the right in the left hand; that in case of no signal by fingers, then the accost was to be, 'What news?' which was to be answered by, 'A ship coming,' then 'What ship?' answer, 'A store ship'; that another signal was 'Are all friends there?' answer, 'Yes, till death.'* Evidence was taken as to certain convicts who had and had not recognised these signs. Another witness said there were three hundred convicts ready to rise, and they would be led by Joseph Holt, the Irish rebel leader who had been transported but was allowed to live and work as a free man. Holt was interviewed and said he was Protestant and had no intimacy with the Catholics.

One can imagine the potency of this combination of English paranoia and Irish dreaming, hatred and loquacity. The committee concluded that there had indeed been seditious meetings held by *several of the disaffected Irish convicts, tending to excite a spirit of discontent which was fast ripening to a serious revolt.* It recommended that five of the Irish receive five hundred lashes, four other *banditti* receive a hundred, and that they and a number of others be sent to *a distant and remote part of the territory*, which meant Norfolk Island.

King formed volunteer groups of fifty armed men at Sydney and Parramatta, to be known as the Loyal Associations or the Volunteer Associations. William Balmain commanded the former, Richard Atkins the latter. These would assemble to assist the army should an uprising occur. Only civil officers, householders and other propertied civilians could enrol. Further, the order stated that *no prisoner or free man (who is not a settler) is to leave the place where he is stationed or resides without a pass from the nearest magistrate.* Settlers were immune: as in Britain, the possession of property was assumed to say a great deal about where one's loyalty lay.

On 26 September 1800, Macarthur, in his capacity as the commanding officer at Parramatta, received a note from 'G.W.'. It was probably delivered to him at home. Written in a trembling hand, it advised that *the infatuated people have planned a general rising tomorrow night and that men from Sydney are to come up here* to join in. The plan was to surround the church and rush in, killing the officers and other gentlemen, at which it was hoped the ordinary people would join the rebels. Macarthur immediately informed King and suggested they do nothing in the hope of luring the rebels out into the open. After sending his message to the governor, Macarthur rushed in to Parramatta to find it in uproar. John Lewis, another of those in the know, had also informed to the authorities, and two of the conspirators had been arrested. Macarthur put the soldiers on alert, hoping that the other rebels might try to rescue their comrades.

Magistrates Atkins and Marsden interrogated some suspects and tortured one in an attempt to extract information. *He behaved so very insolent, and appeared so very hardened*, recalled Marsden, *that Mr Atkins and I ordered him to be punished very severely in hopes of making him inform where the pikes were. Though a young man, he would have died upon the spot before he would tell a single sentence. He was taken down three times – punished [i.e. flogged] upon his back, and also on his bottom when he could receive no more on his back. [He] was just in the same mood when taken to the hospital as he was when first tied up ... I am sure he will die before he will reveal anything of this business.*

A group of officers examined the evidence gathered from various sources – this time some pikes were actually found – and recommended that five convicts receive a thousand lashes, four more five hundred, and another seven two hundred each. The ringleaders were then put into chain gangs on the public works. King lamented that in neither of the two planned uprisings was there sufficient evidence to hang anybody. Dissent continued: in November a group of convicts seized the sloop *Norfolk*. The next month an uprising on Norfolk Island was foiled; thirty pikes were found. A leader on this occasion was one of the Irish recently sent from Sydney because of his involvement in the planned uprisings there. At Christmas another uprising was supposedly uncovered at Parramatta, but it too fizzled out.

On 28 September 1800, King had sent a most comprehensive letter to the Duke of Portland outlining the financial situation and prospects of the colony. He noted that *the introduction of a breed of Spanish sheep into the flocks of individuals has so much improved the fleeces that there is a promising appearance of a great quantity of wool being produced in a few years*. He thought this would be used for the production of clothing in the colony. On the same day he wrote to Sir Joseph Banks, enclosing a large box of fleeces from John Macarthur's flock and asking for an opinion of them. King deserves some of the credit for the development of the Australian wool industry. Indeed, it was he who had persuaded Captain Henry Waterhouse to bring the original sheep from Cape Town in 1797. It is interesting to see that the authorities in England had a firm view of the colony's place in the wool industry even in its infancy. When King proudly told Portland's successor, Lord Hobart, in August 1801 that 306 yards of blanketing had been made from local wool, Hobart responded that the best wool must be exported *for the market of this country* (i.e. England). Cloth production in the colony, he said, must be confined to *the coarser kind.*

In December Judge Advocate Richard Dore died and was replaced by Richard Atkins. Writing to Under-Secretary King, Governor King said Atkins was *the only man in the colony equal to the business, of which he acquits himself very well, and I hope will continue to do so, although I am sorry to say he is sometimes addicted to the bottle ... It is doubtful to me if a fitter man can be sent from England for the present salary*. For all his faults, Atkins was the colony's most socially distinguished inhabitant. Indeed, his wife, Elizabeth, came from so far up the scale she refused to socialise with other women in New South Wales.

On the last day of the year, King issued an order referring to the spirit of disorder still abroad in the colony. *Several worthless vagrants have gone into the woods*, he noted, subsisting by *depredations on the industrious*. He urged the convicts and emancipists to *reflect on the comforts and blessings they enjoy and may acquire in this colony, by a very moderate industry*.

1801

In January a practice alarm was given to test the colony's readiness to respond to an uprising, for the Irish kept coming. A few weeks later, the *Ann* arrived from Cork, *with 135 of the most desperate and diabolical characters that could be selected throughout the kingdom, together with a Catholic priest of most notorious, seditious, and rebellious principles*. In fact, they had tried to seize the vessel off the coast of Brazil. King reminded Portland that the colony now held six hundred United Irishmen who had sworn (back in Ireland) to defeat the British. He said he could handle the situation, but wondered if the colony might have a break from the inflow of Irish rebels. Both Paterson and he had asked for the New South Wales Corps to be strengthened, but to no avail. On 27 May the news came that England had unified with Ireland. King was delighted, hoping that this would reduce the Irish desire for rebellion. Cannon were fired around the harbour to celebrate the occasion, and at Dawes Point the Union Jack as we know it – St Patrick's cross having been added to the previous flag – was flown for the first time in New South Wales. Also in this month, an order was issued permitting people to fire on Aborigines at Parramatta and elsewhere to drive them away from habitations. Two whites had recently been speared.

In June Portland replied to King's request that he purchase the Macarthurs' farm and stock. He wrote that only the English cattle and the Spanish sheep should be taken, as they would improve the breed of the public stock. Everything else – the other animals and the farm itself – would be more useful to the settlement if sold to private buyers. In parentheses Portland wrote that, as Macarthur was a serving officer, *I can by no means account for his being a farmer to the extent he appears to be, and I must highly disapprove of the commanding officer of the Corps to which he belongs allowing him or any other officer to continue in such contradictory situations and characters.* Portland had always been aware that farming was carried on by the officers; presumably it was the detailed description of Macarthur's possessions, and the price tag of £4,000, that brought on this outburst. However, now his attention was captured, Portland had Under-Secretary John King write to Major-General Robert Brownrigg at the War Office along the same lines, and Brownrigg replied that Paterson had been sent a letter expressing *His Royal Highness's entire concurrence in the sentiments of His Grace*, and commanding that officers not be permitted to engage in farming. This order seems to have disappeared, as no more is known of it.

On 12 June the transport ship *Earl Cornwallis* made its way up the harbour, carrying some melancholy news for the New South Wales Corps. A new officer, Lieutenant Henry Crawford, had fallen overboard at Rio de Janeiro and drowned. Paterson was doing some exploration of the Hunter River at the time, so Macarthur was in command at Sydney. It was his responsibility to arrange for Crawford's possessions to be sold and the proceeds sent to his next of kin. He believed that the ship's agent, a naval officer named Lieutenant James Marshall, should have secured Crawford's possessions. But when asked, *He astonished me: he knew nothing of them.* This struck Macarthur as irregular, but he said nothing and obtained the items from the ship's captain. An auction was held, and some of those there later told Macarthur that Marshall had substituted an inferior gun of his own for Crawford's, and swapped a counterpane for a horse rug. Macarthur raised this with Marshall, who at first denied the charges. Macarthur suggested that the easiest way to resolve the issue would be if Marshall produced his weapon so that his accusers could verify whose it was. *This staggered him*, Macarthur subsequently claimed. *He faltered, hesitated, and at last said he would not be positive; the guns might have been exchanged by mistake; if so it should be rectified*. He also admitted he had swapped the counterpane for the horse rug, to his own advantage, but that this was done because Crawford had owed him some money. Macarthur was now convinced of his guilt, and told him to return the items. If it could be proved Crawford had owed him money, this would be paid to him.

Marshall was examined by a bench of magistrates who found some exculpatory evidence. His servant gave information in his favour and it emerged that Marshall had told the quartermaster of the regiment of the proposed substitution. King gave Marshall a severe reprimand *for allowing any part of his conduct giving reason for suspicion* and the matter was closed, at least officially. But Marshall was unhappy, and later confronted Macarthur on the parade ground. *Where's my gun?* he demanded. When Macarthur said he didn't know, Marshall said, *It's a lie! Like all the other lies you've told me!* Macarthur shook his swagger stick and said, *Now, now. If your intentions are to insult me, you have effected your purpose.* Marshall responded: *Don't you shake your stick at me, you ging rascal, or I'll knock you down.* Macarthur went in search of his friend, Captain Edward Abbott, who agreed to be his second.

Abbott delivered Macarthur's written challenge to Marshall, who was staying with Isaac Nichols, he whom Macarthur had tried to frame some years earlier. Marshall accepted, and nominated a man named Jefferies, the purser of the *Earl Cornwallis*, as his own second. Macarthur called on King and told him what was going on. It was a private conversation: of course the governor ought not to know about the duel officially, but the two men were on friendly terms, and perhaps Macarthur was trying to cover his back.

Then the affair took an unusual turn. Marshall and Jefferies went to the place chosen for the duel but the others did not turn up. What had happened was that Abbott had discovered Jefferies was selling goods from the ship behind a counter in Sydney. This meant he was not the captain's social equal; therefore, as far as Abbott was concerned, the duel could not take place. (Abbott was the son of a cook, therefore perhaps sensitive regarding his own claim to the status of gentleman.) Marshall took his refusal as a grievous insult, and went looking for the two men. Near the barracks he found Abbott and walked up to him. Abbott said he would not even discuss the matter, so Marshall replied he was *a damned scoundrel*, and belted him on the side with his stick. Macarthur appeared and Marshall approached him in a threatening manner, abusing him as he went. Macarthur drew his sword and said he'd run Marshall through if he didn't back off. Soldiers came running and the incident was over.

It is curious that Abbott did not feel obliged to challenge Marshall for the blow. Again, this was tricky. In 1746 in Britain a soldier named Lieutenant Ferguson had been struck by an Ensign Campbell, who'd been court-martialled. In that case George II had been so perturbed that he'd actually told the officers' superior that unless Ferguson called Campbell out for having hit him, his honour would be ruined and other officers would refuse to serve with him. And as we have seen, in 1790 a marines ensign in New South Wales was cashiered for not responding to an insult with a challenge. But in the present case, Abbott's fellow officers seem to have decided Marshall's behaviour showed he was not a gentleman, which relieved Abbott of the need to challenge him.

When King heard of these incidents he ordered that Marshall stand trial for assault and battery on Abbott and assault on Macarthur. The trials are notable for several reasons, not least because they were possibly the first criminal proceedings against an officer or any other gentleman in New South Wales, whose trial system at the time was among the most primitive in the British Empire. While this was all right where convicts and the lower orders more generally were concerned, would a gentleman accept it? To make matters worse, of the seven officers on the bench, a majority were members of the corps, hardly expected to be objective in judging a man charged with assaulting two of their colleagues. These officers were William Paterson, John Piper, Neil Mackellar, Thomas Davies and Hugh Piper.

The other members of the court were Judge Advocate Atkins, and Lieutenant James Grant, a naval officer. When the trials began, Marshall asked if he might challenge any of the officers, just as an accused in Britain might challenge a juror. He was told he could not. He tried to ask some questions and was told to be quiet. During one of the trials the prosecutor whispered continually to one of the witnesses. One of the members of the court interrupted the prosecutor at times to ask his own questions and comment on the answers. All up, it was – although normal by Sydney standards – a strange experience of justice for Marshall. But he defended himself stoutly, responding to Macarthur's prosecution by saying: *I cannot avoid expressing my astonishment at the effrontery of Captain Macarthur in my presence, in the presence of any man of honour, attempting to justify his late conduct; he would fain persuade you that I am the coward; but you know perfectly well from your own and from the world's knowledge of that man, and the circumstances which have appeared before you, as well as those upon record in the history of this colony which of the two shrunk from his engagement – I will say from his claim, if a man so acting has any claim, to honour.*

Fine words, but the men of honour convicted Marshall on both charges. For assaulting Macarthur he received a special verdict (to be determined by the secretary of state) and for the attack on Abbott he was fined £50 and sentenced to twelve months in prison. It was the first time in Australian history that a gentleman had been sentenced to gaol.

These court cases are notable because of what they reveal about duelling and honour. Abbott and Macarthur were responsible for their own prosecutions, and Marshall for his own defence. Issues of honour were canvassed at some length because it was felt, rightly or wrongly, that Marshall's assault on Abbott and attempted assault on Macarthur might be partly excused if he had been denied the rightful opportunity to fight the duel. This provides us with one of the best records we have of the lead-up to a duel, including the importance of language, manners and fine points regarding status.

Abbott recalled that Macarthur had come to him, after Marshall accused him of lying about the gun, *very much hurt at the insult that had been offered* and asked him if *I would as your friend [i.e. second] take a letter to him [Marshall], which letter contained a challenge.* The question of the language used by the two parties was important to them. Macarthur stressed that Marshall had called him a *rascal.* Abbott stressed that the letter, requiring that Marshall provide Macarthur with *satisfaction as a gentleman in one hour's time*, contained *no language unbecoming to a gentleman.*

Abbott had then called on Marshall, who was at Isaac Nichols' house. When Abbott entered the room, he took off his hat. However, Marshall did not rise from the sofa. Marshall *received the letter from me sitting* and, after reading it, said he would *meet Captain Macarthur and the whole Corps.* This was not the way a gentleman was supposed to greet another gentleman coming into his room; nor did this response to the letter display the seriousness expected in an affair of honour. Abbott said he was *extremely exasperated at the unhandsome manner he received me.* Marshall – who later claimed he thought Abbott was a servant until he'd read the letter – then stood up and said he needed more time *to get his friend*, and it was agreed that the duel would take place in four hours, at 5.00pm.

Abbott left, and half an hour later was approached by Jefferies, the young man he had seen working in a shop, who announced he was Marshall's friend. To this Abbott *objected, in as handsome manner as I could, Mr Jefferies' going out with Lieutenant Marshall as my co-equal.* Jefferies went off to tell his principal, and at 3.00pm Abbott received a letter from Marshall asking him to name two gentlemen who might act as his second. As a visitor, Marshall presumably had a limited acquaintance among the colony's gentlemen. Abbott replied that he *felt awkward in naming his friend, but to show him that I was not willing to start any further objection, proposed Captain Tennant as a gentleman I could have no objection to and who came out in the same ship with him; premising that it was not my business to point out who his friend ought to be.* According to Abbott, Marshall then sent back a message saying Abbott's letter had been impertinent, and that he would let the matter rest until he met Macarthur again. It seems Marshall had decided to proceed with Jefferies as his second in the hope this would force Macarthur and Abbott to attend. But they didn't.

At 9.30pm Abbott received a letter from Jefferies asking why Macarthur and he had not turned up for the duel. Abbott told the court: *I felt myself very much hurt at this liberty on the part of Mr Jefferies, whom I had before told I could not have any correspondence with as the friend of Lieutenant Marshall considering him according to the rules of duelling not my co-equal.* Macarthur told the court that, after Abbott had refused to act as his friend if Jefferies was involved, he had asked almost every other officer in the corps, and all had likewise refused. Macarthur and Abbott were now anxious that no one accuse Macarthur of cowardice for not fighting. So it was necessary to demonstrate that Abbott had not been overly discriminating in refusing to act if Jefferies was involved.

Later in his evidence Abbott admitted, under questioning by Marshall, that he had discussed the duel with other military officers, who had said *that Captain Macarthur has been too hasty in calling out Lieutenant Marshall, whose conduct had been such as not to be considered by them as worthy to be called out by him.* Marshall, still trying to show Macarthur was a coward, suggested to Abbott that it was interesting timing that the officers should have formed this view so soon *after Captain Macarthur had refused to meet me.* He also, pointedly, asked why Abbott had agreed to carry Macarthur's challenge to someone *so unworthily thought of* as himself. To which Abbott virtuously replied, *I confess this is the only part of my conduct that I have reason to find fault with.*

Marshall continued to pursue the subject of Macarthur's valour, and therefore his honour. He asked Lieutenant Hobby if he was acquainted with *the nature of duelling.* Hobby said he was not, but acknowledged after another question that he had once been *employed in a business of that nature.* Marshall asked him if he'd ever heard of a duel being declined because of the rank of one of the seconds; Hobby said he hadn't. Later, in his final address to the court, Marshall said: *Gentlemen, the laws of duelling, as Captain Abbott has termed them have been brought onto the carpet, in the forlorn hope that a meaning might be tortured out of them to justify the cowardice of the prosecutor … [but] it must be in the recollection of you gentlemen, that his royal highness the Duke of York attended the call of an inferior officer.* Of Macarthur's refusal to fight him he concluded, *This is the first time I ever heard seconds in an affair of honour were to compare their ranks with as much precision, as they measured the lengths of their principals' swords.* He apologised for taking up the court's time on such matters, but it was *done in vindication of my honour, much dearer to me than existence.*

Macarthur argued that it was Marshall who had *avoided* the duel *by declining to nominate an accredited gentleman as his friend.* He offered the court Marshall's letter in which – according to

Macarthur – the writer said he was dropping the matter. And he said that the letters would show *which is the officer and the man of honour, and which the man of rudeness*. There the question of honour ended. Marshall, as we have seen, was found guilty of assault.

Whereas Macarthur's duel with Thomas Gilbert in 1790 had been a simple matter, the Marshall incident, like the Kellow–Faddy one on Norfolk Island, shows how complicated an affair of honour might become. Questions of definition were vital. Were those involved entitled to be so? Did the insult warrant a duel? Such questions were to be answered by the consensus of one's peers. Perhaps this – the need for flexibility and local knowledge – was why the code of honour was never codified, for to do so would have made it overly rigid and prescriptive.

Marshall protested to King, in his role as the only avenue of appeal, about the irregularities of the court proceedings. King, while not doubting his guilt, ordered the court to reconvene to consider his complaints *not only to defend its own integrity, but to afford the most unequivocal justice to the prisoner, and to enable me to judge how far he was deserving of any part of the royal mercy*. The court met to consider King's order and declined to obey it, on the grounds that the governor was seeking to interfere in the process of justice. This decision, involving men of stature such as William Paterson and Neil Mackellar, the governor's aide-de-camp, was a watershed in the relationship between King and the military. King told them they had publicly insulted him and he would be referring the matter to the Home Secretary. The military officers refused to back down and asked King to give London a copy of their letter, so it would be known they had acted as they had only to avoid dishonour to themselves.

On 18 August Macarthur wrote to King saying, with

regard to the report being prepared for London, that *the most extreme caution and circumspection will be necessary on my part to prevent my conduct being misunderstood, and to secure it from the effect of misrepresentation.* It is not clear exactly why he was so concerned, but he went into detail about how King had been aware of his actions towards Marshall from the start, and had approved them. King responded to this letter with an equally detailed account of just what he had and had not been aware of, and had and had not agreed with. He was at pains to make it clear he had learned of the proposed duel (*of how far you had gone on a point of honour*) in a private conversation. This correspondence suggests that Macarthur was warming up and King was wary. Presumably Macarthur, as a colleague of most of the members of the court, was upset by King's attempt to interfere in the results of the trial, and may also have resented seeing Marshall escape prison.

One problem here was that two principles had collided. The first was an Englishman's right to a fair trial, which Marshall clearly had not received, whether he was guilty or not. The second was the independence of English courts from political interference. King had tried to intervene, even if asking a court to review its own decision was just about the mildest form of intervention imaginable. But because the governor combined the executive power with some of the judicial one, such interventions could always be interpreted as political and resented accordingly.

Now Macarthur's antagonism towards King exploded, with a fury impossible to account for from what we know. He began to gather evidence from other officers as to exactly who had said what to whom regarding points of the affair that had no bearing on its central issue. A correspondence with King ensued. The dispute took on a life of its own, with every letter from King containing, in Macarthur's mind, more insults that required more responses. King himself in his replies to Macarthur seemed to have dug in, perhaps wary lest London consider he had been too close with his friend at the time of the proposed duel. Possibly some of the drawbacks of the local legal arrangements were more apparent to him now than they had been. Nevertheless, King comes across as a reasonable and fair man trying to sort out a mess. Macarthur emerges from this correspondence as someone with a considerable capacity for egotistical anger. This was demonstrated by his next action: believing he had been grievously insulted, he put out the word that the governor was to be sent to Coventry; no military officer was to dine at Government House or otherwise visit it except on official business. Here we see two of the problems of the code of honour: its capacity to make disputes that arose in the line of duty personal, and then escalate rather than resolve them.

One searches for some reason for the depth of Macarthur's anger. A possibility is that he was upset by King's crackdown on his trading and other economic activities, which had begun the previous year. More recently, King had helped small settlers delay repaying debts owed to the officers, and opened the government shop in Sydney, thereby making *every poor man respect me*. King himself felt that he had thereby totally suppressed the trade of *the late commissioned hucksters*. Some observers have even found in Macarthur's actions here evidence of early stages of the insanity that was to claim him three decades later. It's also possible his anger arose solely from the questions about his honour raised by the Marshall episode.

In a private letter to Under-Secretary King of 21 August, Governor King noted that his response to Marshall's trial *has finally sent me to Coventry, where I have no sort of objection to remaining for the time His Majesty may think proper to continue me in the command of the colony, which I have not a wish to extend a moment beyond the time of my conduct being approved of.* He had been in the job less than a year. Reviewing the Marshall case, he predicted that *the arts and intrigues of a man you have heard so much about (I mean Captain Macarthur) will one day or other set this colony in a flame.* He noted how Macarthur had managed in the Marshall case to *involve me in their [the corps'] dirty dispute*, and that Paterson is *a weak, honest man, and if he acted from his own ideas of right and wrong would do the right thing; but, unfortunately, he is like many other men, 'made a knave's tool of'.*

Macarthur now tried to drive a wedge between King and Paterson, who to his chagrin continued to be on friendly terms. He claimed grimly that all the officers had agreed on a policy of social exclusion towards the governor and that Paterson had reneged. In early September he told Surgeon James Thomson, a magistrate, that Paterson, despite the *resolution of cutting with the governor*, had frequently been seen since at the governor's, *not on points of duty, as he was in coloured clothes*, behaviour Macarthur termed *very pusillanimous.* At some point in this unfolding drama, King invited all the officers apart from Macarthur to dinner at Government House to test their loyalty. All attended except four: John Piper and his younger brother, Hugh, Edward Abbott, and William Minchin. Macarthur was losing control of the situation, which seemed to enrage him further. He claimed to some that Paterson had actually initiated the boycott before betraying it. As a result of this, Macarthur was determined to let the governor know *what the colonel had expressed, both in correspondence and opinion, respecting the governor.* This was a reference to a letter Paterson had recently sent his patron, Sir Joseph Banks, asserting that King *has carried his economy too far*, and claiming that he had missed an opportunity to buy grain. The letter was drafted with the help of Macarthur; indeed, King later wrote that it had been *extorted* from the colonel by *the reports and persuasions of those who were aiming at anarchy and confusion.*

Paterson had sent another letter, on 23 August, to General Brownrigg at the War Office, protesting against King's attempt to have the court reconsider the Marshall case. He requested that the New South Wales Corps officers should never again *be threatened in the manner they have been for discharging their duty according to their conscience in a court of justice, where they are sworn to decide in obedience to the laws and not in conformity to the pleasure of the governor.* The threat referred to was a charge of disobeying the governor, which King was sending to the Home Secretary. Paterson then set out the nub of the matter: *most of the disquiet which has agitated this settlement, and does agitate it, is chiefly to be attributed to the unfortunate mixture of civil and military duties which exist in this country*. He asked that officers no longer be required to sit in court, and begged that the government depose Richard Atkins, the judge advocate, *from the office he so unworthily filled, and which it is astonishing he was ever thought by Governor King a proper person to appoint to, as his character for low debauchery and every degrading vice, as well as a total want of every gentlemanly principle, is universally known throughout this settlement.* King later said, perhaps rightly, that Macarthur had had a hand in this letter too.

Now Paterson, threatened with exposure by Macarthur, approached King and told him of the existence of these letters. Rumours of them were already circulating among the officers. Soon King asked Paterson for copies. He persuaded Paterson his criticisms were wrong, and forgave him for acting *as the instrument of one who had a certain point of resentment to obtain*. King asked Paterson to dinner. The colonel accepted. Macarthur was told and proceeded to up the ante even higher, and publicised a letter from Paterson's wife to his own that put its writer in a bad light. Macarthur also told others of conversations that had occurred when the Macarthurs were staying in the Patersons' house in Sydney. (Just what Elizabeth Macarthur thought of this is not recorded.) The insult to Paterson contained in these disclosures was considerable. As in the dispute with King, one notes how a relatively modest disagreement, based on professional duties, was accelerated by Macarthur into full-blown personal warfare.

At last Paterson responded to these deeply offensive breaches of confidence in the only way possible: he demanded a duel. His second was Captain Neil Mackellar, who agreed that Macarthur had acted in *violation of what has ever been held sacred amongst men of honour.* On 10 September he called on Macarthur and said: *As you have abused the confidence Colonel Paterson had reposed in you, he expects you will give him that satisfaction he, as an injured man, has a right to require.* To this Macarthur replied, *Whenever he pleases.* His friend Captain John Piper agreed to be his second.

Macarthur's behaviour in breaking confidences was highly dishonourable, and a hundred years later it's possible he would have been shunned by society for such offensive behaviour. But in his time it was expected that Paterson, the man offended, challenge the person who'd insulted him to a duel. This of course was grossly unfair, as it put the offended man in danger. It also gave the dishonourable man a way out, as the duel, once over, was generally considered the end of the matter. As this shows, the code of honour was not about honour in the sense we would use the word. Nor was it about determining who was right or wrong in any situation. Rather, it was a way of limiting aggression by bringing matters to a head.

Paterson's decision to call out Macarthur was later criticised by King. Some believed it was better, where an officer and his subordinate were to fight, to wait until both men were off active duty. In the 1770s the governor of Madras, Lord

Macartney, had been upset by one Major-General James Stuart, but had waited until they were both back in England to fight him. (Stuart had only one leg and had to be leaned against a tree; Macartney was short-sighted. The affair ended without bloodshed.) But by betraying the confidences of their wives' friendship, Macarthur had given Paterson little alternative. They had to live side-by-side in a small community indefinitely. It was behaviour well beyond that acceptable from a gentleman, and there was no redress to be obtained under law, civil or military. There was also the equally pressing problem of Macarthur's turning some of his fellow officers against their commander. A duel was needed quickly, to clear the air.

Duelling between officers was contrary to the army's Articles of War, section 7, article 2, which declared *No officer … shall presume to send a challenge to any other officer … to fight a duel, upon pain … of being cashiered* (i.e. dismissed from the army). Moreover, according to article 3 of the same section, *all seconds, promoters, and carriers of challenges in order to duels, shall be deemed as principals, and be punished accordingly*. This placed everyone involved in a difficult situation. As the anonymous author of *Cautions and Advices to Officers of the Army* wrote in 1795, *such is the unaccountable prevalence of custom, that the disobeyer of these orders is generally applauded, while the obeyer of them is obliged to quit the army with disgrace!*

No one involved in this case had any intention of obeying these orders. The duel was set for the next day, but bad weather postponed it until 1.00pm on Monday, 14 September 1801. The venue was switched from Sydney to Parramatta, although the precise location is unknown. Surgeon John Harris, who had been Macarthur's second at the duel at Plymouth back in 1789, attended in his capacity of doctor. Macarthur's pistols were faulty so he loaded them himself, which was not usually done. (Later several officers examined the guns and agreed that, due to various defects, no one unacquainted with one of them could load it *without the greatest danger of shooting himself.*) Mackellar loaded Paterson's guns. The seconds then tossed a coin to see who would shoot first, and Piper won. They measured out twelve steps, and the principals took their places facing each other. Macarthur raised his gun and fired, hitting Paterson – who presumably was standing side-on with his weapon down by his side – in the right shoulder, and the colonel dropped his pistol. Harris rushed up to him and inspected the wound. There was still a chance that Paterson might return fire, but the injury proved too serious to allow this. Mackellar later said that Macarthur then *sent me a message, as if exulting in victory*, something that would have been extremely insulting and dishonourable, and which Macarthur strenuously denied. Everyone went home.

The ball could not be extracted from Paterson's shoulder and the doctors couldn't say if he would survive. Although the British army tolerated duelling, despite the Articles of War, its attitude changed if anyone was killed. This was rare, but it did happen. Earlier that year in Bombay a Lieutenant George Bellasis of the 16th Regiment had killed a man in a duel. For this he was sentenced to fourteen years' transportation to New South Wales, where he arrived in 1802. However, he was freed immediately and joined the New South Wales Corps – but he had not shot his commanding officer.

At the same time as the duel was occurring, King left Sydney for the Hawkesbury. Stopping at Parramatta, he learned of what had happened and at 8.00pm put Mackellar and Piper under close arrest in their barracks. (It's not clear why Macarthur didn't receive the same treatment.) The two captains had their swords taken from them. The next day Adjutant Minchin took the seconds to Government House, where King interviewed them and ordered them to provide written reports of the incident on the spot. It was an understood role of seconds to tell the world how a duel had proceeded, particularly whether the protocols had been observed and how the principals had behaved. If anyone died and there was a court case, the jury would be influenced by whether the duel had been conducted properly, in accordance with the code of honour. King particularly asked Piper to explain why he had cooperated in a duel involving his commanding officer; but Piper's report was short to the point of contempt, and made no response to this request. When the reports were done, the governor released Mackellar because he was needed for his job of managing public works. His sword was returned to him while a fuming Piper was taken back to the barracks, still under arrest. King was upset that he had *gone out* against his commanding officer, *a man who had always acted towards him as a father and a friend*. He also considered it a departure from *the rules of propriety* that Piper had allowed Macarthur to load his own pistols, and reprimanded the captain severely. One reason this mattered was that it was the seconds' responsibility to see that all the pistols were loaded properly, for the sake of fairness.

Perhaps, as Paterson lies on his bed poised between life and death, this is an appropriate point at which to offer some reflections on John Macarthur, his world, and violence. By now the themes in his life are clear. He was a man with considerable intelligence and an ambition that was both aided and undermined by an unusually deep, reckless and imaginative capacity for aggression. Later he was to observe some of his own characteristics in his son John: *an indescribable fierceness of independence and an obstinacy to pursue what he has once determined on, which neither reason nor dread of future consequences are likely to operate on him to relax*. He had clawed his way up to become a gentleman landowner and at times seemed nervous about the security of this status. He was still ambitious. New South Wales was not a sufficiently large or complex stage for a person of his considerable abilities, and he considered leaving it several times. But in the end he always rejected the difficulties and uncertainties of returning to England. Possibly this made him more restless and aggressive than he might have been had his energies been provided with greater challenges. Certainly, there is something desperate in the effort and creativity he devoted to his disputes.

Macarthur and his contemporaries called themselves *men of honour*, and the phrase deserves the mafia connotations we bring to it today. Like *cosa nostra*, the New South Wales Corps adopted the code of honour to fill a power vacuum created by a weak state. Yet this was changing. The power of government was increasing, and men like Wilberforce were creating a new culture in which duelling would one day be unnecessary. But that lay some decades in the future. Meanwhile, it was a society whose uncertainties encouraged Macarthur's form of fluid, intelligent and enthusiastic aggression. He could switch between the law and the duel with a spontaneity denied men whose sense of order and morality was more pedestrian. He used both methods as means not of achieving social harmony but of triumphing over anyone who opposed him.

In 1911 the German economist and sociologist Werner Sombart published an essay arguing that there were two types in European history, the hero and the trader. For him, the Germans were heroes while the British were traders, drowning in an *oozing flood of commercialism*. Heroes, of course, lived by honour (duelling continued in Germany until the twentieth century) rather than the rule of law, which was preferred by the traders. But the British had once been heroes, and it is central to Macarthur's life that he, like many of his contemporaries, contained both types within the one disturbed and violent individual.

Macarthur kept up Piper's spirits by telling his friend *a few days' confinement to your room will improve your complexion … I observed to you yesterday that your nose was sunburnt.* But the letter ended on a harder note: *You will, of course, be careful to avoid making any common cause with people that I cannot act with or for.*

The whole affair was deeply troubling to King, for he realised that his old friend Paterson, in challenging Macarthur, was more to blame than anyone. At this point he received the surgeons' report and learned that Paterson's life was still in danger, so he put Mackellar back under arrest, and arrested Macarthur. Deciding that it would not be proper for officers to be placed *in a common jail among the worst of felons*, he put the participants under house arrest until it was known whether Paterson would live. Looking ahead, King also announced that Macarthur was to be transferred to Norfolk Island. The next day Macarthur wrote to the governor in protest, pointing out that it was usual for a more junior officer than himself to be stationed there. King referred him to the original order, where he had said that the transfer was necessary for *the good of His Majesty's service in this colony*.

Macarthur wrote to Piper again, congratulating him on the attitude of hostility he was showing King and offering to give him money should he be sent to Britain under arrest. The letter provides an insight into Macarthur's thoroughness when he had committed himself to a quarrel, and is worth quoting at some length. He began with patronising flattery: *Your letter greatly rejoices me because it displays … that your honour, your faith and your character are of more consequence to you than any other consideration.* After chiding his friend for not providing more details in his report on the duel for King, he said he was attaching a questionnaire on the event for Piper to complete and return. As always, Macarthur was aware of the need to prepare the right documents. He then explained: *It is absolutely necessary for both our characters' sake that we never speak to Mackellar again, for if we did, it would look as if we admitted the truth of his most rancorous statement [i.e. that Macarthur had exulted after shooting Paterson]. All our friends are of that opinion and are determined to stick to it, except Abbott, who I have had no opportunity to consult, but who I am certain will see the necessity of it and readily come to the same resolution.* He continued to the mild-mannered Piper: *Let me however charge you not to think of laying a hand upon him [Mackellar], for he is my game, and he who interrupts me in the chase will be regardless of the law of friendship. Abbott of course has shown you my letters to the governor – do not care for me – for I am now so deeply in that the game begins to be amusing.*

What a game. What a strange man.

Even though Macarthur was confined to Elizabeth Farm, King was taking no chances, and ordered Mackellar on no account to fight him. Were Mackellar killed, the world would lose his evidence of Macarthur's allegedly shameful behaviour at the end of the duel. King must have relished the prospect of Macarthur's standing being seriously damaged if it could be proved he had exulted over a wounded opponent.

On 19 September it was Piper's turn to sting the governor with a letter. He complained that King had reproached him *in a manner and in language very hurtful to my feelings*. He had been kept under close arrest, and demanded to know why he, the second, was receiving *more rigorous usage* than Macarthur himself. He also asked for a copy of the charges on which his arrest was based. King replied, merely referring him, like Macarthur, to his previous orders. Two days later, Piper wrote again to explain, in response to King's earlier chastisement, that he had agreed to be Macarthur's second because they were friends and he considered Macarthur the injured party. Venturing onto the tricky moral terrain of just what degree of gratitude he as a *gentleman* ought to feel for Paterson, he concluded that supporting his fellow captain's *honour* superseded the debt he owed his lieutenant-colonel. He then turned to the vexed question of why he had let Macarthur load his own pistols, saying *His pistols are so defective about the locks that no person but himself can load one of them without danger, and the other requires a very correct management in half-cocking or it will miss the full-cock.* Macarthur had conveyed these technical details to Mackellar in Piper's presence before the duel, *observing at the same time that he made a point of honour to make it known least Colonel Paterson should request one of his pistols, and it should deceive him* (i.e. misfire). Mackellar had replied that *he knew one pistol was defective, but that Colonel Paterson would use his own – a new case of valuable ones.*

The Macarthurs continued to work on Piper to make sure he remained hostile to King and didn't crack. John arranged it so the two of them should send an insultingly identical letter to King refusing to promise (as had been requested) that they would not break the peace if released. Again he offered to pay Piper's debts if he was sent home: *no man shall call after you for a shilling – are we not suffering in one cause and ought we to have more than one interest?* He signed this letter, *Your much obliged and ever faithful friend.* Elizabeth sent her sympathy and urged Piper *to keep up your spirits and be circumspect and cool in what you do – remember the old adage that when things are the worst they oft times mend.*

Now three officers had been under arrest without being charged for eight days, the maximum allowed under military law. King ordered that they be released on bail, on condition they would appear in court should Paterson die. Macarthur, who had been fuming at his treatment, refused to have his arrest withdrawn. It seems strange to us that he could actually do this, but he did, offering to continue to perform his duties so long as he could be court-martialled in order to clear his name. King's order putting him under arrest had *publicly disgraced* him, he wrote, and *appears as if intended to prejudice the world against my conduct, and deeply to wound and injure my reputation as an officer and a gentleman.* He felt that *the good of His Majesty's service* required that *no officer should submit to unmerited disgrace and to injury without seeking for a public inquiry, and, if innocent, for public redress.* Regarding the various claims in circulation about his own behaviour, he said *I have betrayed no private correspondence, no private conversations, ... I have displayed no exultation over a wounded opponent, or in any way behaved unlike a gentleman ... on the contrary, I am the person who has been betrayed, who has been exulted over, and who has been treated with the basest ingratitude and the blackest treachery.*

These comments are central to our understanding of Macarthur's behaviour after the duel. They were substantially untrue. We know from his behaviour in other disputes that he could defend himself ably if the evidence was there to support his point of view, going into considerable detail. That he did not do so in this case makes it likely that he had in fact betrayed Paterson's correspondence and conversation, as Paterson and King claimed he had. The super-heated tone of the letter also suggests its writer was protesting too much, although admittedly Macarthur had a liking for hyperbole even when in the right. Given Macarthur's considerable capacity for self-righteousness, it's possible that, although he must have been lying about the betrayals of correspondence and conversation (for why else would Paterson have called him out?), he believed that King's action in arresting and failing to charge him blotted out any faults he himself might have committed. But Macarthur's defence was presumptuous, arrogant and potentially reckless. He was a mere captain, a young man with no supporters in England; King was an older, experienced and senior officer respected by his superiors and, like Paterson, friendly with Sir Joseph Banks. To take on such men was dangerous. King decided it would be too difficult to try Macarthur in the colony, so he returned him to house arrest and ordered him to prepare himself for a return to England, where he would be tried for attempting to create dissension between King and Paterson.

Maintaining the pressure, Macarthur shot back the request: *I also wish to be acquainted whether my arrest is intended to be so rigid as to prevent me from walking over my own grounds for the benefit of exercise and health, as a close confinement previous to the voyage I have to undertake must necessarily make me very unfit to embark on it.* King told him he could walk in his garden but no further. Then the governor began to assemble information to send back to the Duke of Portland that would assist him in understanding John Macarthur and what he represented. He set out to prove that Macarthur had sought to drive a wedge between Paterson and himself, and provided character assessments from others as attachments to his own long letter. Richard Atkins wrote that he had no information about recent events, but Macarthur's *infamous and diabolical conduct, his rapacity in accumulating a large fortune in so short a time, his extortions on the industrious and laborious settler, which has plunged themselves and [their] families in distress and misery … [have] considerably impeded the happiness and prosperity of this colony.* Moreover, *unless he is sent Home as a common disturber of the public peace, and as a man who has most essentially injured H.M. service, this colony can never enjoy that happiness and prosperity it is H.M. wish that all his subjects should be partakers of.*

Surgeon James Thomson said that Macarthur had told him

the officers of the New South Wales Corps, including Paterson, had agreed to avoid the governor socially because of the Marshall affair, and that Paterson had later reneged on the agreement. Moreover, Macarthur had told Thomson that he had been induced to cut King by Paterson and another officer, who had both now deserted him *and left him supported only by a very few*, and because of this *he would assuredly let the governor know what the colonel had expressed, both in correspondence and opinion, respecting the governor.* However, Thomson went on to say that *Colonel Paterson has frequently assured me that he never had assented to such resolution as alluded to by Captain Macarthur, but, on the contrary, had refused when urged to it by Captain Macarthur.* Paterson, who was now back from death's door, confirmed this, adding that other officers, including Mackellar, had heard him talk about it at the time.

Several people, including Samuel Marsden, supported Paterson's claim that Macarthur had threatened to reveal his private criticisms of King to the governor unless he boycotted private functions at Government House. Paterson now told King he thought a court martial of Macarthur for insubordination (assuming the right charge could have been found) might have failed, as some of the officers who would have sat on it supported Macarthur rather than himself. Thus his only recourse had been to call out Macarthur.

On 6 October Macarthur decided to throw a party for the soldiers under his command, possibly as a present to them before his departure. The spirits that were to be drunk on the occasion were seized by the constables while they were being moved without a permit the night before, much to the anger of the troops. King informed Paterson that he greatly disapproved of an officer in Macarthur's situation giving his troops a dinner and spirits, and ordered that he give them no more drink.

King now finished his letter to Portland, suggesting that Macarthur had advised the officers who'd sat on the trial of Lieutenant Marshall and then, when King had dared request the court reconsider its decision, had provoked members of the corps to break with the governor. As to why King did not court-martial Macarthur for creating dissension in the colony, it was because King himself was so involved in the case, as indeed were most of the officers who could have sat on the trial, that there could be no hope of an objective outcome. King decided that Piper could be court-martialled in Sydney, however, and Mackellar would go to England, bearing Macarthur's sword and a copy of the governor's charges relating *to the restless and turbulent conduct* of its owner, to which he would be able to give witness.

To assist his case against Macarthur, King wrote Under-Secretary John King a private letter giving more background on Macarthur, who he claimed had arrived in Sydney in 1790 £500 in debt and was now worth £20,000. This was almost certainly a gross exaggeration, as Macarthur had offered to sell his house, land and stock to the government the year before for only £4,000. King claimed Macarthur had sent enormous sums of money back to Britain, but there is no evidence of this. He went on to say that Macarthur was now wealthy enough *to boast of his indifference of whatever change happens to him*. This was probably true, and we have already seen how irritated King was that some of the officers were so rich that he couldn't hurt them. His fury rising, he wrote: *Experience has convinced every man in this colony that there are no resources which art, cunning, impudence, and a pair of baselisk eyes can afford that he does not put in practice to obtain any point he undertakes*. He was, King summarised, a *perturbator*, and a *master worker of the puppets he has set in motion*. Macarthur was to receive more criticism in his life, but this letter contains the most famous invective he ever attracted. The *basilisk* (correct spelling) is a mythical creature, hatched by a serpent from a cock's egg, whose look can be fatal. A *perturbator*, as the word implies, is a disturber or troubler.

King claimed that Hunter had challenged Macarthur to a duel, and that Macarthur had declined, adding that he himself would have challenged him but for the adverse effects this would have had on the colony. These revelations are interesting. They indicate the acceptability of duelling, as King was prepared to disclose this information to the under-secretary, albeit privately. They also show Macarthur's extraordinary capacity to get under the skin of anyone unfortunate enough to be placed in a position of authority over him. King's reluctance to challenge Macarthur was understandable: in 1780 during a war with Mysore its governor-general, Warren Hastings, had fought a duel and been criticised for it because of the implications for British rule in India had he been killed.

In concluding, King noted that Macarthur had been the *arbiter of the colony* before he arrived and reduced his power: *I shall close the subject by observing that if Captain Macarthur returns here in any official character it should be that of governor, as one-half the colony already belongs to him, and it will not be long before he gets the other half.*

The letters that passed between all the men involved in this conflict read curiously today. They seem far more bitchy, and obsessed with personal slights, than anything most of us are familiar with in official correspondence. One of the reasons for this was probably sheer fear. Life at that time was arbitrary. Usually it was not what the world might think of an official that mattered, but what certain important people and their advisers thought. And such people were back in England, their opinions reliant on what they heard, and on letters. Hence the obsessive concern of all the early governors, and some of the other colonists, for what men such as Sir Joseph Banks and the Duke of Portland might have been told about them, no matter how unlikely. King, despite being a secure and competent man, revealed in his correspondence a great concern with what people were saying about him in London. As Macarthur remained unpunished for having gone over Hunter's head in his letter to the home secretary regarding the Atkins' affair, King's concerns are understandable. Letters were dangerous.

Another reason for the obsessiveness of correspondence was that the concept of honour was relatively subjective. Later in the nineteenth century, particularly as the public school system grew, standards of good behaviour became more clearly defined, and society was more prepared to condemn those who breached them. But in Macarthur's time, while honour was highly valued, what it actually meant in practice was often determined by what you could get away with – or what your enemies could get away with in besmirching it. So smart officials such as King were always keen to ensure the correct interpretation was placed on events.

The above letter was written on 8 November, a week before Macarthur left. He had continued to perturbate, in October buying 1,770 acres near Toongabbie and some 1,300 sheep from Major Joseph Foveaux. The sheep had been on offer to the government, and Macarthur's purchase infuriated King, who tried to get the animals back. But their new owner had already mixed them up with his own, so this was not practicable. Before the purchase, which cost Macarthur £2,000 for the stock and the farm, the vendor had been the biggest, and the purchaser one of the biggest, stock-owners in the colony. The combination of the two flocks meant Macarthur now towered above anyone else in terms of his land and animals.

It is interesting to note that Foveaux had been one of the

leading traders during the 1790s and had received more land grants than most. If his farm and stock were worth only £2,000 in 1801, it suggests Macarthur might have had the best of the deal had he been able to persuade the government to pay him double that for his own, smaller, farms and flocks. It also reinforces the view that King's estimate of his wealth was exaggerated.

At this point Macarthur had not yet received a final response to his offer to sell up to the government. But we can assume, from the expansion of his holdings and the decision to leave most of his family behind when he sailed for England, that Elizabeth and he had decided to stay. It's true he now offered King Foveaux's stock as well, but on new, inflated, terms for the whole deal. King, estimating the purchase would cost the government at least £20,000, recommended against it, telling Under-Secretary King: *This is one of the great monopolist's traits. By his securing such an abundance of sheep he has secured the wool.* He mentioned all this, he said, *to prevent government being imposed upon by this rich Botany Bay perturbator.*

In many of King's criticisms, Macarthur's wealth is a leitmotiv. King clearly brooded on it, and it grew enormous in his imagination. This was understandable: the governor had few opportunities such as Macarthur's, and, like Hunter before him, was an honest man worried about money and the security of his job and, in King's case, his family. A lot of the resentment shown by several governors to Macarthur reflects the combination of envy and distaste with which public servants sometimes regard people in private enterprise.

On 15 November 1801, Macarthur went to sea once again, on the *Hunter*. He was to go home by way of Calcutta, and was taking two of his children to be educated, the eight-year-old Elizabeth and John junior, aged seven. They were accompanied by a servant. Macarthur had been in New South Wales for ten and a half years and was returning to England a fairly wealthy man, yet a relatively powerless one. He had made some dangerous enemies, and had no contacts of importance in Britain: the future must have looked uncertain.

We don't know if the idea of exporting wool was at the front of his mind. He must have been wondering about the future of his large farms, what might be grown on them and where the markets were. He had, after all, sent the box of fleeces to Sir Joseph Banks (via King) some time before. There was money in whales and seals, but trading income was not what it had been. We can surmise that, at the very least, Macarthur was going back to England restless and looking for new opportunities. From Norfolk Island, where the ship called, he wrote to his friend John Piper, *the devil take him [Marsden], King, Paterson and Mackellar. Now, as you are an observing man, see if you can discover such a set in quadruplicate amongst your acquaintance here and elsewhere.*

Captain Neil Mackellar, carrying Macarthur's sword and copies of all the relevant documents (King sent two other sets by different routes) left Sydney on the American whaler *Caroline* on 29 March 1802. Effectively he was to be Macarthur's prosecutor in England. But neither he nor the *Caroline* was ever heard of again. This affair, which had begun with one death – of the officer whose gun Lieutenant Marshall was supposed to have stolen – ended with another. The unfortunate Mackellar's death was a considerable stroke of luck for Macarthur, and he was to receive an even greater one shortly.

VI

Luck

Macarthur was returning to England via India: by 1801 some shipping companies had tacked Sydney on to the end of their Asian routes. But early in the voyage the *Hunter* was struck by a typhoon in the Arafura Sea and dismasted, so the captain put in at the port of Amboyna (today's Ambon, in Indonesia), controlled by the East India Company. This was a former Dutch trading post, a ramshackle place dominated by a fort and filled with Malays, Chinese and some Europeans. The man in charge was a young East India Company resident, Robert Farquhar, who had recently attacked and captured a nearby Dutch outpost. Fighting between European merchants, often using local soldiers, had been frequent in Asia for many years, but in this case Farquhar's superior considered he'd been disobedient and demoted him. Macarthur told him that if he accepted the demotion *you will be laid on the shelf for life. Tell them they are not fully aware of the position of the Dutch settlement, or the circumstances which led to your attacking it, that you are convinced of the necessity for taking it.* And resign.

It sounds like Macarthur's policy: never apologise; never take a backwards step; raise the stakes. Farquhar followed this advice, resigning and sending a strong letter of self-defence to Lord Wellesley, the governor-general in India. Wellesley was impressed by this response and took Farquhar under his wing, promising him the governorship at Penang. Macarthur sailed for England with a letter from Robert to his father, expressing his debt to the visitor from New South Wales.

And this is where the Macarthur family's fortune changed, for Robert's father was Sir Walter Farquhar, friend and physician-in-ordinary to the Prince of Wales. Sir Walter, according to Macarthur family tradition, received the visitor *on the most friendly and confidential footing, making his house and family, as it were, a home for Macarthur and his children.* At last Macarthur had the one thing he'd always lacked: a powerful patron. The importance of this was to be enormous. Alan Atkinson has shown that Farquhar was a leading member of a group of powerful Whig families who had become wealthy as the empire grew. These families included the Stevensons and Hallidays. Their patron was the Prince of Wales and their opponent the King, with his Tory supporters in parliament. Farquhar and his friends and relatives had no problem accepting Sydney's richest man into their circle. They could see how he might help them in the future, just as they could help him.

However, not everyone in London welcomed Macarthur. After leaving Ambon and then India, he had stopped at the island of St Helena, where he wrote to Piper that *every person acquainted with our story applauds my conduct and execrates Mr King's and that of his venal associates.* There he was entrusted with some natural history specimens from Africa for Sir Joseph Banks. Their safe delivery, when Macarthur and his children finally reached London in late 1802, was greeted with *a dry note of acknowledgment.* Macarthur was offended, but perhaps he should have been worried. The Tory Banks was still the government's main adviser on New South Wales and Macarthur had argued with two of Banks' protégés and shot one of them.

Macarthur was gratified to have the matter for which he'd been sent to Britain resolved in his favour. According to the army's Adjutant-General Harry Calvert, Governor King had *judged ill in sending Captain Macarthur away from the spot where the transactions took place.* This was because, due in part to the absence of the unfortunate Mackellar, there were no witnesses in Britain apart from Macarthur himself, which made a court martial unfeasible. Macarthur was censured for his *erroneous notion that an officer who has been put under arrest has a right, as it is termed, to demand a court martial upon himself, and may persist in considering himself as still under the restraint of such arrest, although expressly released therefrom by the superior officer who imposed it.* Still, Calvert was not too harsh, adding that the notion was widespread, *even amongst some intelligent officers.* On 1 February 1802 a general order was issued to the British army to ensure that never again would British officers act as King and Macarthur had in this case. Thus Macarthur was responsible for one of the turning points in the code of honour. No longer could military gentlemen manipulate the law by insisting on a pointless trial to restore their honour after charges had been dropped. From then on, they could be arrested and later freed, just like everyone else.

Of more immediate importance, though, was the fact that Calvert took Macarthur out of arrest and advised that on his return to New South Wales, *His Royal Highness conceives that Governor King will not be desirous of bringing the articles of complaint, whatever they may be, which he has against Captain Macarthur, before a military tribunal.* We don't know what role, if any, Macarthur's new friends played in this splendid outcome of his dispute with King. William Balmain, who had returned to London to live, wrote sourly to D'Arcy Wentworth: *Macarthur has been relieved from his arrest ... but in every other aspect he has got the worst of his contention whatever he may write to the contrary. Vast indeed is the difference he finds between this place and New South Wales, we are all very small folks here.*

But some were no longer as small as they looked.

Sydney

Back in New South Wales, John Piper was tried for his part in the duel. Elizabeth urged him to *avoid all offensive in your defence. It is useless, my good friend, to add fuel to the fire that has been blazing too long already*. She asked him to send her a copy of what he proposed to say at the trial, which reinforces the impression that Piper's relationship with the Macarthurs was like that of a younger brother. He was found not guilty.

In June 1802 the two ships of Nicholas Baudin's expedition arrived at Sydney, and remained there for five months. Napoleon had sent them to investigate the claim by an American captain that he had sailed up a strait through the middle of the continent. If this was the case, Napoleon hoped that the western part could be claimed for France. It was also a spying expedition. The French had long been interested in the Pacific. De Bougainville had crossed the ocean some thirty years earlier. La Pérouse, of course, had reached Botany Bay just after the First Fleet arrived, and when he later disappeared D'Entrecasteaux was sent to find him. French translations had been promptly published of the First Fleet books by Watkin Tench, Surgeon John White and Arthur Phillip. It was not just the Pacific that fascinated the French but the success of the colony of convicts on its rim, which they regarded as an important social experiment. In 1789 parliamentarian Pierre-Edouard Lemontey had predicted poetically in Paris: *Perhaps a strong and hard-working nation will emerge from a vile rabble of convicts, as in other times a swarm of ruffians founded the empire of the Caesars. Life everywhere is born of corruption. It is to fetid dung that we owe both the golden harvests and the dazzling vine.*

Now was the chance for the French to evaluate the great experiment first-hand. Baudin was impressed by what he found: *I could not regard without admiration the immense work that the English have done [at] ... Port Jackson. [It is] difficult to conceive how they have so speedily attained to the state of splendour and comfort in which they now find themselves.* He wrote to the minister of the navy: *I should warn you that the colony of Port Jackson well merits the attention of the government ... People in France and elsewhere are far from being able to imagine how large and prosperous the English have been able to make this colony in the space of fourteen years.*

Baudin was to die on the way home, and the report of the expedition would be written by naturalist Francois Péron, who recalled *how astonished we were at the flourishing state of this singular and distant colony.* When Péron visited Marsden's Mamre Farm beyond Parramatta he noted: *No longer ago than 1794, the whole of this spot was covered with immense and useless forests of eucalyptus ... and it was over a very excellent road, in a very elegant chaise, that Mr Marsden drove me to it. What pains, what exertions must have been taken to open such communications! – and these communications, these pastures, these fields, these harvests, these orchards, these flocks, are the work of eight years.*

These reflections need to be discounted. France was so backward in comparison to Britain it's probable they reflect the writer's first contact with rural British society in general, rather than its manifestation in New South Wales in particular. In France in the eighteenth century, at least until the Revolution, most of the population was starving and almost half of all children died by the age of ten from malnutrition and disease. For people from this background, exposure to British prosperity, even in its Sydney version, must have been shocking. Still, the accounts of the French are useful in providing a contemporary outsider's view. Of the convicts, Péron observed that *they have been forced to lay aside their anti-social behaviour. The majority, having atoned for their crimes by a hard bondage, have rejoined the ranks of the citizens*. This picture is a long way from the Bacchanalian anarchy portrayed by people such as Hunter, King, and Marsden.

The colony continued to grow in extent and complexity. In 1802 the first book printed in Australia, a collection of government orders and notices dating back to Governor Phillip, came off the press in Government House in a print run of eight hundred. Also this year, the *Buffalo* arrived with some mares plus a large stallion sent as a present from the Duke of Northumberland to Major George Johnston. It was nearly seventeen hands high, and towered above all other horses in Sydney. Northumberland was the patron of Johnston, who had returned to Sydney after the charges against him for paying a soldier with spirits were dropped.

In late 1802, relations between King and the military deteriorated again. There was a complex matter involving an allegation that King had allowed some of Baudin's officers to purchase spirits from a ship in the harbour and sell them on shore, while the officers of the corps were forbidden from doing this. The claim was wrong, but it became lost in a chain of counter-claims and general abuse, and there were court-martials and further complications. King became upset on learning that Paterson had apparently known of the allegation but had not told him of it. Paterson too became upset. Three months earlier, he had received an order from the commander-in-chief of the army that officers in the corps were not *permitted on any account whatever to engage into the cultivation of farms, or in any occupation that are to detach them from their military duty*. Now he informed King of this, and asked that several officers who'd been assisting the governor be returned to the corps. King obliged, but was outraged that Paterson continued to allow the officers to farm. King sent back to Paterson, unasked, his bodyguard of five soldiers and replaced them with emancipists. The next time there was trouble from the Irish, King sent this bodyguard to assist, but the corps refused to serve alongside them. This led to more furious correspondence between King and Paterson, and more problems. In January 1803 Lieutenant Thomas Hobby obtained a pipe critical of the governor and showed it to his colleagues before it was passed to Government House. The delay in handing over the offending item so outraged King he ordered Hobby and another officer be charged.

On the day he received this order, Paterson had himself examined by no fewer than five doctors, all of who concurred that he needed to maintain *as easy and tranquil a state as possible*. This clearly ruled out any leadership position in New South Wales, and Paterson informed King that, on doctors' orders, he was now withdrawing *from all business* until his health was improved. This seems to have been what we would call stress leave, for later he wrote to Sir Joseph Banks explaining how, following the falling-out with King, *Not having a very good constitution, the anxiety and uneasiness I suffered injured my health so materially as to require absolute retirement*. He had also, of course, been shot by John Macarthur.

George Johnston took over as commander of the corps and continued the running argument with King, who finished one letter: *I beg to intimate how convenient it would be to the contents of my writing case if our correspondence could be abridged, if not cease, as I fear you will soon oblige me to write on brown paper, of which I have but a very small stock.* All this indicates that Macarthur was not the only military officer capable of infuriating the governors, nor the only man in the colony who could turn a small dispute into a large one.

Finally, King snapped. In May 1803 he wrote to Lord Hobart in a similar vein to a letter his predecessor had once sent to London. So rattled was he by the continuing battles with the New South Wales Corps that he virtually apologised for not challenging its officers to duels (he had been prevented only by *the high sense I have of the important trust delegated to me*) and demanded either an official inquiry into his administration or permission to visit London to justify his conduct. At which one wonders: justify what, and to whom? Whereas Hunter's paranoia was firmly rooted in his administrative incompetence, King was a much better governor, with many reforms to his credit, and had received the gratitude (as he himself assured Hobart) of all classes of people in the colony other than the military. He was an intelligent man who had been prepared for resistance and calumny even before he left Britain. So why did he break at this point? We know he was sick: in May he wrote to Banks that *I am frequently attacked with our old friend the gout, which I think is turning into an acute rheumatism … at times I am disabled from going out for a fortnight.* It's possible that after three years he was just worn down, by his illness, the isolation of his position and the loss of former friends such as Paterson and Macarthur. He had probably realised by now the impossibility of achieving any more reform, given the strength and influence of the officers of the corps, the inadequacies of the legal system, and the indifference of the government back home.

The fights never seemed to stop. King was in the middle

of a furious row with a naval captain, James Colnett of the *Glatton*, who had asked him to pardon a young convict woman transported on his ship, with whom Colnett seems to have formed a relationship. He also asked King for a land grant. In each case King quite properly refused, which led Colnett to abuse King and then, on his return to London, write blistering and contemptuous letters of complaint to the Admiralty and to Lord Hobart (whom he also visited) in which he condemned every aspect of the governor and the colony. We might think King should have been able to shrug off this sort of thing, and trust to the judgement of his superiors. But for whatever reason, his troubles had got on top of him. Hence the letter asking that he be allowed to visit London. He probably needed a holiday as much as anything.

On 5 March 1803 the first issue of Australia's first newspaper, the *Sydney Gazette*, appeared. It was a four-page weekly and ex-convict printer George Howe depended for his income on advertising and private print jobs. Occasionally publication would be shortened or even suspended when the colony ran out of paper. The newspaper carried official information, shipping and other news, private notices, advertising, and literary work, including poetry by Howe himself. Each issue had to be read and approved by an official before publication. Surveying the early copies, Sir Joseph Banks noted they were *rather under the influence of government* but praised the venture and asked to be sent every issue.

1803, London

Britain's clothing manufacturers were seeking industrial relations reform. They wanted a relaxation of the laws governing their employment of artisans, and it was important for their case, for reasons that need not detain us here, to demonstrate to the government that their industry was capable of expanding. To prove this they had to show that vast supplies of raw materials, specifically fine wool, were available. But they had a problem – supplies were actually dwindling. Due to food shortages caused by the war with France, British farmers were increasingly growing sheep for mutton, not wool. And Spain, the world's leading producer of fine wool, had come under French influence and supplies were drying up. (Eventually France was to invade Spain.) As the manufacture of cloth was a major industry, this matter had acquired political importance at the time of Macarthur's arrival in London.

The manufacturers appointed two representatives, Thomas Atkinson and A.L. Edridge, to attend the progress of the industrial relations reform legislation through parliament. These men called on Macarthur, who had brought some wool with him. He later recalled: *I immediately gave them an order for its delivery from the India Warehouse, and they obtained the case containing the fleeces and took it to their committee room. The wool being examined was highly approved and the different kinds valued.* The men asked Macarthur about New South Wales' potential as a source, and he described its recent past and waxed lyrical about its golden future as a producer of fine wool. They were ready to believe him – they were desperate and it was an incredibly optimistic age. Historian John Gascoigne has quoted the view of an economist of the period: *No man can pretend to set bounds to the progress that may yet be made in agriculture and manufacture … is it not much more natural and reasonable to suppose, that we are rather at the beginning only, and just got within the threshold, than that we are arrived at the ne plus ultra of useful discoveries.* John Macarthur certainly felt no desire to set limits. However, there were two things necessary to achieve this vision: the right type of sheep, and land. Lots of land.

Macarthur claimed that he himself had started with hairy sheep from India (so-called Bengals) but in 1797 some merinos had been imported from the Cape of Good Hope by Captains Waterhouse and Kent. These, or rather their ancestors some years before, had come to South Africa from Spain via Saxony and Holland. Macarthur said that he, unlike other farmers in the colony, had kept the merinos separate from his other sheep, and as a result he now had better quality wool than anyone else in New South Wales. He modestly attributed considerable foresight to himself in focusing on wool as far back as 1797 because he had realised it might be *of the utmost consequence to this country*.

Atkinson and Edridge were impressed by some of Macarthur's wool, and by the story he had to tell. In July he even published a pamphlet on sheep and New South Wales, and the manufacturers' representatives reported that the colony could *with proper attention* increase its supply so that the price of wool would drop and Britain would become independent of Spain. They urged other manufacturers to write to the government *both as Britons and Commercial Men*, urging it to encourage the growth of wool in New South Wales. When these letters were received, the Office for Trade in Whitehall naturally turned to Sir Joseph Banks for advice. As it happened, Banks was a respected expert on sheep as well as on New South Wales, and had helped bring merinos to the royal flock. He was told that Macarthur had predicted that *millions of [sheep] ...may be raised in a few years* in New South Wales at very little expense. Sir Joseph proceeded to scoff at these claims, informing the office that New South Wales was unsuited to growing wool. *I have never heard of any luxuriant pastures of the natural growth of New South Wales at all fitted for the pasturage of sheep*, he sneered, *till I read of them in Captain Macarthur's statement*. He also alleged that freight costs from Sydney to Britain would make the proposal uneconomical and urged that no special encouragement be given to Macarthur. Overall, he entertained *no manner of hope* that New South Wales would prove to be a good place to raise sheep. This harsh rebuttal presumably stemmed from a combination of factors, including Macarthur's puffery, and Banks' experience of the *tall coarse and reedy* grass found at Botany Bay. Most important of all, perhaps, were his friendships with King and Paterson. After this, nothing more happened for a while.

We know few details of how Macarthur spent these years in Britain. He devoted much time to his children, seeing Edward during the holidays, and placing John in school with him. When he decided the English weather did not suit Elizabeth, he found a governess prepared to return to Sydney with them. He would have spent time cultivating the Farquhars and their allies, and otherwise pushing his interests in London. It's possible – likely, knowing Macarthur – that he used his new Whig contacts to agitate for King's removal. In October two letters critical of King's administration appeared in the *Morning Post*; King blamed them on Macarthur. In November Macarthur wrote to John Piper calling King *the tyrant* and said he himself was *up to the ears in papers for carrying on the war against our common enemy*. Now that Banks was opposing his interests, Macarthur had another reason to want King removed.

In November Macarthur went to the church of St Giles-in-the-Field in London for the funeral of William Balmain, the man whose challenge to a duel he had once declined. As winter set in and there was no progress in his affairs, Macarthur decided to leave for Sydney in February. But the government faltered, in the face of a coalition of the Whigs led by Charles Fox and Lord Grenville's party. Given his new patrons, Macarthur hoped for great things should the Whigs obtain government, and decided to stay. In January he wrote to the Treasury, proposing that a public company be established to grow wool in New South Wales. It's strange that he was now asking for a public company – admittedly one to be run by him – rather than more direct assistance for himself. Maybe he wanted to share the risk; maybe he had reason to think such an approach had the greater chance of success.

March 1804, New South Wales

The colony continued to be a violent place, one where the worst criminals were left hanging from the gibbet after they had been executed, to be seen by everyone who might pass, free and convict, adult and child. Aggression was still common. In his memoir of his years in Sydney, David Mann recalled the murder in August 1803 of Joseph Luken, a constable, by villains *who buried the hilt of his own cutlass very deeply in his head.* Mann was second on the scene and, telling us perhaps more detail than we need to know, described how, in attempting to turn the corpse, *my forefinger penetrated through a hole in the skull, into the brains of the deceased.* Mann informed his readers that in his history he was recording only the more gruesome events, as he had not the space to describe the less sensational murders. Around this time a woman named Salmon killed her new-born child by throwing it down the privy, where it was smothered. The mother was sick and died before she could be punished, and was buried beneath the gallows. In September they tried to hang Joseph Samuels, a burglar, but the rope broke. They tried again but the knot slipped loose. So they strung him up a third time, and amazingly the rope snapped again. At this point the provost marshal rode off to the governor to inform him of what had happened and Samuels was given a reprieve. He later drowned while trying to escape from the colony.

But the main fear of the colony's establishment continued to be rebellion. On the evening of 4 March 1804, Elizabeth Macarthur and her children dined at the Marsdens' Parramatta residence. She had not seen her husband for several years now, and was running their extensive farms herself. Matthew Flinders had observed the previous year that she was *torn with anxious suspense and borne down with the cares attendant upon the interests of a large family*. Governor King was good to her; indeed, as the invitation to the Marsdens suggests, her social life might have been easier because her husband was on the other side of the world.

While they were eating, a bell rang out at the government farm at Castle Hill. At the same time one of the convicts set fire to a small hut there. These were the signals for the beginning of an uprising by many of the Irish convicts and other disgruntled folk. It may have been inspired by the news, which had reached the colony six weeks earlier, of a rebellion in Ireland the previous year. At Castle Hill, two hundred convicts broke into buildings, armed themselves, and then rampaged through the countryside, stealing more guns and supplies from the homes of small settlers. As a girl had been raped during a recent outbreak of fourteen convicts, women and children now fled into the darkness to escape the approaching mob. In fact the rebels were surprisingly well-behaved, apart from stealing 180 more swords and guns. One settler, William Joyce, the manager of the Macarthur farm at Seven Hills, managed to escape and rode into Parramatta, which he reached at 9.00pm.

The first Elizabeth knew of the uprising was when a servant burst into the parlour *pale and in violent agitation. Sir, says he, looking wildly at Mr Marsden, come with me. And you too Madam, looking at me. Then half shutting the door he told us that the croppies had risen*. The Marsden and Macarthur families rushed to the barracks. The fire at Castle Hill could be seen from Parramatta, and the Irish were expected to enter the town at any moment. Captain Edward Abbott, the commanding officer, sent a messenger to Sydney and called up the Loyal Association of armed volunteers, who were joined by other settlers. Meanwhile, those escaping from the mob poured into the town.

It was the rebels' intention to pause, hidden, outside Parramatta, while a small group of confederates set fire to Elizabeth Farm to draw the garrison away from the town. At that point allies in Parramatta would rise up and the rebels would rush in to join them. At 11.00pm the Marsdens, the Macarthurs, Abbott's wife and one other free woman took a boat to Sydney, which they reached at 3.00am. They found the place in an uproar, and were helped ashore by officers from the naval vessel *Calcutta*, which sat in the harbour with its lights blazing. There were lights and soldiers everywhere, and a roadblock had been placed on the western road. Elizabeth and the children were escorted to the Marsdens' town residence.

The governor had already left on horseback for Parramatta, followed by fifty-six members of the corps on foot. At Annandale Major George Johnston was roused from bed at midnight, and took charge of this detachment when it reached his house at 1.30am. He checked that the men had good flints in their guns, which he ordered to be loaded. The march continued and they reached Parramatta at 5.00am. King gave Johnston written instructions which, due to the poor handwriting, he couldn't make out. *I told him no, he might as well have given me Greek.* The governor then read the orders to Johnston, who, on hearing that he had permission to fire on any rebel who tried to escape, told him that was all he wanted. During the night the rebels' attack on Parramatta had not occurred. No one had tried to set fire to Elizabeth Farm, and nothing was heard from any confederates in Parramatta itself, which was now heavily guarded. So the rebels, numbering some 233 men, had moved off in the direction of the

Hawkesbury, where they hoped to find more recruits. The colony's radicals had always believed a mass uprising would occur if only some brave souls would act first. Now this belief was to be tested.

The day dawned very hot. After some brisk marching, the troops caught up with the rebels at 11.00am, on a small rise about five miles from Toongabbie along the road to Green Hills. With Johnston were some twenty-nine soldiers, sixty-seven civilians, mostly armed members of the Loyal Association, one trooper and Father James Dixon. The priest's background provides a glimpse of the passionate English–Irish enmity that lay behind this rebellion. He had been transported after the 1798 uprising in Ireland, simply for being related to a rebel leader named Thomas Dixon. (Admittedly, Dixon had captured ninety-seven government troops and had them piked to death and thrown off a bridge.) Another of Dixon's relatives had been hanged by the government at Wexford Bridge.

The trooper with the soldiers was called Thomas Anlezark. Johnston sent him ahead to lie to the croppies, telling them that the governor was coming to negotiate, in the hope of making them stop so the rest of the troops could catch up. When this didn't work, Father Dixon was sent to persuade the rebels to surrender, but to no avail. Then Johnston asked Anlezark if he would ride up to the mob with him, and the trooper replied *he would go to hell with me*. The two galloped up to the croppies and used deceit to capture two of their leaders. By these ruses the soldiers had had time to arrive, and they proceeded to fire on the mob and then charged them. The rebels, who had been calling for *liberty or death* all night, broke and ran. Johnston's men were so enthusiastic in their pursuit that he had to threaten to shoot some of them to stop them killing the convicts they had taken. Altogether, about fifteen rebels were killed and six were wounded.

No one from the other side was hurt.

So ended the most successful convict uprising in Australian history. The site of the conflict became known as Vinegar Hill. One notes, apart from the general incompetence of the rebels, their good behaviour as they raided farms for arms, the fact there were English as well as Irish in the mob, and the lack of stomach they showed for a fight as soon as the action started. Given the fears the establishment had harboured for so many years, the uprising, when it came, was anti-climactic. It is conventional to introduce an adjective like *farcical* at this point, but it was actually sad, that this was the best the Irish could achieve, with their considerable numbers and all their passion. The rebel leader, Philip Cunningham, who had promised to burn the government store at Green Hills, was hanged that day from its staircase. There was no trial for him. Other ringleaders were dragged before the courts and then hanged or flogged. George Johnston received a grant of 2,000 acres for suppressing the rebellion, and remained in charge of the corps. Later in the year, William Paterson sailed off to establish the first settlement in northern Van Diemen's Land.

Elizabeth Macarthur and her children were soon able to return to their home, where she continued the long wait for news of John. *Week after week passes on and leaves me a prey to the most anxious and cruel uncertainty*, she wrote. *The management of our concerns gets burthensome to me in the extreme, and I am perpetually annoyed by some vexation or other.* The next month, two of the Macarthurs' stockmen were killed by Aborigines on land they owned between Prospect Hill and the Cowpastures. One of the Aborigines involved, named Taloon, was later killed by a party of settlers.

In June 1804 Assistant-Surgeon James Thomson wrote a general account of the colony for Under-Secretary Edward

Cooke. He noted that Robert Campbell and the emancipists Simeon Lord, Henry Kable and James Underwood had built or bought ships and were active in whaling, and that sealing down in Bass Strait was also on the increase. According to Thomson, the main thing lacking in the colony was *a virtuous and industrious peasantry*. Most of the emancipists were a lost cause, mired in *idleness and drunkenness*, while the ex-soldiers who made up a large proportion of the free settlers were not used to hard work and gladly swapped their land grants for spirits. As for the families sent from England as free small settlers, they were often no better than the convicts, drunks and criminals who knew nothing of agriculture and were now fed by the state.

Thomson thought all this particularly regrettable as *In no place in the world might the inhabitants be more happy. Living in a country which is not subject to any contagious diseases, excepting during a few of the summer months, when the putrid dysentery is rather prevalent.* There was plenty of productive land out of Sydney, he thought: *Fruit, potatoes, and vegetables of all kinds are in abundance; poultry is reared in great quantities about every house; and, in fine, everything to make people happy who are not determined to be discontented with everything. Of this some good mechanics who had been convicts were sensible, having plenty of employment, lived with their families in great comfort and even luxury, comparatively speaking, with the manner in which people of the same description live in England.*

March 1804, London

With the increasing prospect of a change of government, Macarthur decided to launch a counterattack against Banks. On 4 March he wrote to Captain Henry Waterhouse, who as well as bringing the first merinos to Sydney had lived and farmed in New South Wales. He responded in complete agreement with Macarthur's view of the colony's fitness for sheep and described the Cowpastures, across the Nepean and at the foot of the Blue Mountains, as *a beautiful park, totally divested of underwood, interspersed with plains, with rich luxuriant grass*. Macarthur then called on Banks, who was out, and possibly left a copy of Waterhouse's opinion for him. A few days later Banks wrote a conciliatory letter to Macarthur. He asked if he and the others to be involved in the proposed public company would be satisfied with one million acres *as sheep-walks only*, in other words as leases rather than freehold. It appears from this suggestion that Banks had seen Waterhouse's letter, and was now trying to broker some compromise offer. Possibly he realised that his reputation as an expert on wool and New South Wales was under threat from Macarthur's determined campaign.

In May the government changed. Overall it was not what Macarthur had hoped for, as the King, who hated Fox, appointed William Pitt as the new prime minister. However, the new Secretary of State for War and the Colonies (now responsible for New South Wales) was Lord Camden, whose private secretary was George Watson, a relative of the Farquhars. This was for Macarthur a most splendid development.

Macarthur wrote to the committee of the Privy Council that had been formed to consider matters relating to trade and foreign plantation. He set out to rebut Banks' arguments against growing wool in New South Wales, and told the same story of a golden future that he had been honing for the past year. (At this time, only about one-tenth of the colony's sheep could be said to produce fine wool.) Banks had asserted that the native grasses could not support sheep, so Macarthur attached the contrary opinion of Waterhouse. Macarthur suggested that, if their Lordships didn't like his idea of a public company, he'd be perfectly happy to shoulder the burden himself if they'd just give him 10,000 acres and thirty convicts as shepherds.

In July the committee interviewed Macarthur, who explained how in New South Wales farmers burned off long grass to encourage new growth on which sheep could feed. He told them he owned about 4,000 acres and one-third of all the sheep in the colony (according to King it was one-quarter) and that no other settler had paid any attention to the improvement of wool (untrue: Samuel Marsden being the main exception). He explained how he housed his sheep under shelter at night because of the wetness of the climate. He offered to proceed with the expansion of wool-growing even if the government gave him more land only on condition that it was a success.

The committee also interviewed a merchant about the cost of transporting wool from New South Wales, and learned that Banks' claims about the exorbitant cost were wrong. Then it was the turn of ex-Governor Hunter, who spoke favourably of wool-growing and of his old enemy, John Macarthur. *I have no doubt*, he said, *that any offer he may make will be worth attending to*. Here was a surprise. Unlikely as it seems, Macarthur had apparently patched up his quarrel with Hunter, and remained friends with him from then on.

All this diplomacy eventually paid off when the Privy Council committee wrote to Lord Camden later in July, observing that New South Wales might be a good place to grow fine wool, and that *a conditional grant of lands of a reasonable extent may be perhaps with safety granted to Mr Macarthur for the pasturage of sheep only*. The recommendation, couched in the language of the public service, was vague and non-committal. But it was enough for George Watson, Macarthur's ally in Camden's office, to go to work with. Presently, Camden announced that Macarthur could sell his commission and have

10,000 acres of the best land in the colony, the Cowpastures.

This was Macarthur's finest hour. He had turned defeat, or at least a potentially dangerous trip to Britain, into a glorious victory, being offered far more land than had ever been given to one person in New South Wales. This had been achieved by taking a few scraps of information, a few fortuitous events and actions of the past five years, and turning them into a golden prospect of a new and enormous national export industry. Having no idea what lay to the west of the Blue Mountains, his vision of the Australian wool industry was fictional. That it turned out to be accurate – even though it took longer to realise than he'd predicted – was pure luck.

Macarthur was not the only wool enthusiast in the colony. Marsden was at least his equal as a breeder, and King is the first person recorded as forecasting a big future for wool. But Macarthur was the one who pulled it all together into a grand scheme: the wool, the markets, and most importantly of all, the land. Yet all this would probably have been for nothing without the other ingredient, his newfound patrons, the Farquhars and their relative George Watson. At the same time as Camden promised Macarthur his 10,000 acres, he gave another 2,000 – still a large grant by Sydney standards – to a nineteen-year-old man named Walter Davidson. Young Walter was Sir Walter Farquhar's nephew.

In August, Macarthur attended the first sale of some of the Spanish merinos from the royal flock, at Kew. These sheep had been obtained illegally from Spain by the agents of Banks, who was also at the sale. Banks, despite his recent, positive letter to Macarthur, proceeded to announce publicly that it was illegal to export sheep from Britain. Macarthur insulted Banks in front of several noblemen, and bought eight of the forty-five beasts on sale. Shortly after this, an article in a newspaper repeated the information that it was illegal under an old law to export sheep from Britain, and that the penalties included forfeiture of the ship and branding of the culprit's hand. Macarthur went to Lord Camden's office and passed Banks coming out. Inside, he was told Banks had been insisting the sheep could not be exported. Macarthur explained that 10,000 acres without sheep was not much use, and after some discussion Camden issued a treasury warrant permitting the export.

Macarthur's problems with Banks were by no means over. A few days later Camden sent for him and said Banks thought 10,000 acres was too much. So the grant was changed to 5,000 acres down and another 5,000 maybe, if Macarthur could produce large quantities of fine wool. In the months to come, after Macarthur's departure, Banks continued to urge that he receive no more land, perhaps egged on by a letter from King referring to the perturbator as an *artful and unprincipled sheep-monger.*

In this affair Banks seems to have been torn between his dislike of Macarthur and his realisation that Macarthur might, however gallingly, be right. Banks knew his sheep, and he knew that the British Empire, while covering an increasingly large area of the globe, contained little good sheep country. Thus his advice to the government varied, his antagonism both moderated by an awareness that no one knew what sort of land lay out of sight in New South Wales, and sharpened by his opposition to Macarthur. Not unexpectedly, Macarthur was enraged by his attitude and would later say in New South Wales that Banks was a *venal and debauched old rascal.*

Finally Macarthur was able to leave, sailing on a ship he had obtained named the *Argo*. There was a large wooden carving of a golden fleece on its prow as a figurehead, and the eight real merinos were securely below deck. He took with him his daughter Elizabeth and her new governess, Penelope

Lucas, his nephew Hannibal Macarthur (the son of James, who had probably lent John the £400 he needed for his start in the army years before), Walter Davidson and several wool classers, tradesmen and servants. Most importantly of all, he carried with him letters from Lord Camden instructing Governor King to give him 5,000 acres and at least thirty convicts to work them.

England, 1805

William Bligh was in trouble again. While the *Argo* made its way around the globe, he was court-martialled in Britain on charges of *tyranny, unofficer-like conduct, and ungentlemanly behaviour.* The charges were those Fletcher Christian might have made had he returned home on the *Bounty*, although otherwise this case bore no resemblance to the earlier one. One of Bligh's lieutenants on his current command, the *Warrior*, was a man named John Frazier, who had injured his leg in an accident. Bligh and he had begun a dispute about just how serious the injury was, and according to Frazier, Bligh did *grossly insult and ill-treat me … by calling me a rascal, scoundrel and shaking his fist in my face.*

The court martial in late February 1805 at Torbay produced a great deal of evidence, some of it contradictory, but there was no doubt Bligh's ability with the language had not deserted him since 1789. He had called the master *a vile man, a shameful man*, the boatswain an *infamous scoundrel* and *a dastardly villain*, the carpenter *a dastardly, cowardly old man.* One of Bligh's terms of abuse was the word *Jesuit*, while a certain gunner was described as *a damned long pelt of a bitch.* In his defence Bligh acknowledged that he might sometimes appear *unnecessarily anxious* for his orders to be carried out, but this was only because of his zeal for the service.

The court found the charges in part proved, and reprimanded Bligh, who was *admonished to be in future more correct in his language.* It was only a tap on the arm, and Bligh returned to the command of the *Warrior.* Writing about the case to his patron Sir Joseph Banks, Bligh abused most of his officers who *were a very bad set of men as I ever heard of ... Robert Cinnamond our surgeon, I really believe is the most designing wicked man ever came into a ship.* As with the *Bounty*, Bligh appeared to have learned absolutely nothing from the experience about himself and his effect on other people.

However, the court martial did so little harm to his reputation that less than a month later Banks wrote offering him the job of governing New South Wales, at twice the salary received by King. While supportive of King, Banks had been critical of his lax standards of punishment. It may have been that recent events had suggested to him that a tougher man was needed to deal with John Macarthur. According to Banks, Bligh was *One who has integrity unimpeached, a mind capable of providing its own resources in difficulties without leaning on others for advice, firm in discipline, civil in deportment and not subject to whimper and whine when severity of discipline is wanted.* This was correct on every point except one: far from being *civil in deportment*, Bligh had the power to infuriate gentlemen over whom he had authority. But it's not an aspect of Bligh's character the aristocratic Banks would have encountered. Bligh got on well enough with those comfortably above or below him on the social scale; it was equals he had difficulty with.

Bligh thought about it. He was concerned that he would not be able to take his wife, *owing to her extreme horror of the sea.* In the end he decided to go after the government agreed to various favourable terms regarding his salary and pension. Bligh had six daughters (all but one of whom he was leaving behind with their mother) and, like King, was worried about money. He swung things so that his son-in-law, Lieutenant John Putland, could come out as a naval commander.

Sydney, 1805

Macarthur's arrival in Sydney Harbour on 7 June was different from his previous entrance, fifteen years before. While he carried with him, on his own ship, a golden future (five rams and the ewe had survived the voyage), he was already the colony's wealthiest man. He had some 3,500 acres, more than twice as much as anyone else except Quartermaster Thomas Laycock (2,115 acres) and Thomas Rowley (1,975 acres). He grew no crops any more but his orchards and gardens extended to 39 acres. (Samuel Marsden, the next biggest gardener, had thirteen.) His flocks numbered 5,320, followed by John Palmer's 1,557 and Marsden's 1,035. In August he had, apart from his employees, thirty convicts.

When Macarthur reached Elizabeth Farm, his five-year-old son William, shocked by his father's pockmarked face, which he thought ugly, burst into tears. Unfortunately we know very little about Macarthur's private life, at this time or almost any other. There are hardly any letters in which he reveals much of himself. His son James recorded, many years later, that John spoke often of *the shining characters of ancient history* and drew inspiration from Romans such as the generals Coriolanus and Scipio Africanus. He was fond of quoting Shakespeare, Milton and Addison. But no sign of these literary enthusiasms is to be found in his correspondence. Some memories seem unreliable. William Macarthur's claim that in boyhood he had only one toy and two sets of clothes is almost as curious as James Macarthur's memory of his father as a man *loth to enter into a quarrel* (although James did add that John was *bold and uncompromising when assailed*).

Macarthur's arrival in Sydney was of course the talk of the town. Referring to the return of *the hero of the fleece*, King told Banks: *After everything that is passed, you may readily conceive that this arrival ... and the manner he was to be received, caused a little sensation.* Macarthur called on King and told him how well he had been treated by Sir Joseph Banks at the Kew sale. Presumably this was intended to impress the governor. Macarthur gave him Lord Camden's letter. After reading it, King offered his hand to Macarthur, *who very gratefully received it ... So much for our meeting after four years of suspense and vicissitude.* Macarthur's own view of his reception was that *the letter Lord Camden had the goodness to write respecting me has operated like a necromantic spell, and lulled every angry passion to sleep. Peace has succeeded ungovernable rage, and those who were ready to annihilate each other are now as friendly in appearance as if their whole lives had been spent in a constant interchange of kind offices.* Warming to his theme (the letter was to an ally in Lord Camden's office) he continued: *Amongst the plants that I have been so fortunate as to bring out alive is the olive, and I hope, as an emblem of peace, its branches and that blessing will spread together, and be universally propagated throughout the whole country.* The olive tree was planted out the front of Elizabeth Farm and is still there.

King had actually been reprimanded by Lord Hobart for sending Macarthur to England for trial, when in the colony alone could *all the necessary witnesses for the prosecution and defence [be] ... found.* King had replied, and we can imagine his teeth gritting as he wrote, that it was not the witnesses but the make-up of the court that was the problem. He also regretted that Captain Neil Mackellar had not made it to London to contradict *such assertions as Captain Macarthur might choose to make for his own advantage, and to have stated such circumstances relating to that officer's conduct as exceeded the bounds of correspondence.* However, King accepted the reprimand *with that respect which a life so far spent and I hope I may add usefully and honorably devoted to His Majesty's service, tells me it is my duty to receive with all becoming humility.* But his spirit was crushed.

On Norfolk Island John Piper had just been appointed lieutenant-governor to replace Joseph Foveaux. Young Elizabeth wrote to their old friend: *I shall venture to inform you of our safe arrival and though late congratulate you on your appointment, notwithstanding that I know that you great people seldom make any other use of your letters than to light your fires. ... You may judge how great must have been our joy on finding my dear mama and all the rest of the family in perfect health ... I have sent you a few seeds which I hope will be acceptable.* Her father also sent a letter, telling Piper that *everything is settled to my satisfaction and that I entertain hopes of universal peace once more resuming her reign in this heretofore unhappy place. My ship shall visit you as soon as she can be got ready for sea and bring you a portion of the comfort we have brought out.* That comfort included a large supply of spirits.

When extra convicts were not forthcoming soon after his return, Macarthur wrote to King explaining that the safety of his flocks, due to lack of manpower, was *highly precarious*. He received fourteen more men immediately. While King was genuinely concerned to assist in the development of a wool industry, he had his own ideas, and suggested to Lord Camden that Macarthur's 5,000 acres be located somewhere other than the Cowpastures, which was needed for the government's wild herd of cattle. King had visited the area in December 1803, when he had had to fend off an enraged bull with his riding whip. Macarthur fired off a letter to Under-Secretary Edward Cooke opposing this and hoping that *the peaceful and productive sheep will find in you an advocate*. One reason he wanted the land at the Cowpastures was that it had high and dry hills behind it, where the sheep might be driven in wet seasons to avoid getting footrot on the soggy lowlands. In fact, Camden is not ideal sheep territory, being too damp; but until the Blue Mountains were crossed it seemed pretty good.

Macarthur and Samuel Marsden were asked by King to compile a report on the colony's experience of sheep-breeding, for which they sent questionnaires to the leading farmers. This information was passed on to Britain. Macarthur then went exploring, for the first time in his life, looking for 5,000 acres on the Nepean as an alternative to the Cowpastures. In October he advised King that the search had failed, as all the country was too wet or covered with scrubby brushwood rather than grass. He asked again for the Cowpastures, confident that *this national object will appear to you of too much importance to be exposed to the risk of failure by confining the sheep to insufficient or unhealthy pastures.* He offered, if allowed to occupy the area, to quit it should Camden support King's reluctance. Perhaps he felt he had minimised the risk of this occurring with his letter to Cooke.

King agreed. In November Macarthur began to move his sheep to the Cowpastures, and on 15 January 1806 he formally received the grant of 5,000 of the best grazing acres so far discovered in the colony. It was an extraordinary coup, and helped ensure his family's prosperity thereafter. Walter Davidson, the living reminder of Macarthur's new influence in London, received his 2,000 acres right next door. Macarthur wrote to King thanking him for the grant, and asking for still more convicts. He said he'd hire free men to do the work but they were unavailable, and finished by asking after Mrs King's health. King wrote back saying Mrs King was fine, and allocating seven more convicts to Macarthur plus a detachment of troops to prevent trouble from the Aborigines.

King seems to have been quite sincere in patching things up with Macarthur after his return. In July he had appointed him Captain Commandant of the Parramatta Loyal Association, and in December he gave 1,440 acres at Cabramatta in grants to Macarthur's nephew Hannibal, his daughter and two employees. Indeed, King had been generous to many people, possibly feeling that Lord Camden's 5,000 acres set a new benchmark. Various officers and the free settler Gregory Blaxland received over 1,000 acres each. King even gave his own children 2,350 acres. In January 1806 he extended by fourteen years the leases he had previously given on land in central Sydney once reserved by Governor Phillip for the Crown. Macarthur's block was next to the absent Colonel Paterson's house and comprised just over one acre. It was given *for the express purpose of erecting buildings upon, suitable to his residence and carrying on his commercial concerns.* Others who had their leases extended were Paterson, George Johnston, Garnham Blaxcell, John Palmer, John Harris and Thomas Jamison.

In mid-1804 King had been informed that he would be replaced as governor because of *the unfortunate differences which have so long subsisted between you and the military officers of the colony.* The Colonial Secretary was disappointed that *the spirit of party … has reached such an alarming height.* This was not quite what King seems to have had in mind when he requested permission to take leave in England to justify his administration, but he accepted with good grace. He was, after all, a sick man, and had given up fighting with the officers by this point, and was seeking an easier existence for himself. On its part, the administration was weary of the disputes between King and the officers. King had actually written to the Home Office that, were Macarthur permitted to return, *my recall, or permission to return, will be absolutely necessary, to prevent such steps being taken by me as will not much tend to the quiet and good of the colony.* One of the meanings of this is that he would call Macarthur out, so his recall might have been prompted partly by a wish to avoid such a possibility. In July 1805, Viscount Castlereagh succeeded Lord Camden as Secretary for War and Colonies (Macarthur had been incredibly fortunate to be in London during Camden's brief term of office) and wrote telling King that *in conformity to your request … the King has been pleased to appoint captain Bligh, of the Royal navy, as your successor.*

1806

In January 1806 King had a conversation with Macarthur, who argued for the free importation of spirits. King replied that as the importation of the small quantities now permitted caused so many evil effects, he would never allow more to come in. The idea that the restriction itself might be responsible for the problems did not seem ever to occur to him. During this talk Macarthur claimed that a lawyer in Britain had suggested that all local regulations in New South Wales, indeed all orders of the governor, were actually illegal. The lawyer Macarthur was referring to was almost certainly Jeremy Bentham, who as we have seen was the advocate of domestic imprisonment for Britain's criminals and a staunch opponent of transportation, and therefore of New South Wales, which he once called *The True Bastille*. In 1803 Bentham had privately published a pamphlet called *A Plea for the Constitution*, setting out what he saw as the illegal basis of the governor of New South Wales' power over free settlers. Bentham noted that the separation of powers, essential to freedom, was completely absent from the colony. He met several times with David Collins, who was soon to leave to try to form a new settlement at Port Phillip. Macarthur, who was also in London at the time, probably heard of Bentham's argument from Collins, or from some other source. Hence his subsequent conversation with King.

In February there was an enormous flood over the low-lying parts of the Hawkesbury settlement, and much grain and many buildings were swept away. The Reverend Samuel Marsden was sent to assess the damage and help with the mopping up. He found that looting was widespread. In March he told King he had had a free labourer flogged because he had refused to assist with the recovery operation. King wrote back approving this action and urged that any other shirkers be flogged too. It is an interesting insight into the legal rights of poor free men (probably in this case an ex-convict) in New South Wales in 1806.

Macarthur wrote to John Piper that *Our friend is at last preparing to take flight and his successor, Governor Bligh, formerly captain of the Bounty, is daily expected.* After describing the recent floods he commented with heavy sarcasm: *What a scene for a new governor and what a fine subject for panegyrick on the care, wisdom and foresight of our friend.* Yet a few months later the Macarthurs were staying at Government House and Macarthur wrote again to Piper that *the governor is uncommonly kind and obliging, in so much that I give you my word I am not very anxious for an immediate change.* Macarthur was later to refer to a proposed deal with King that would have enriched both their families but which did not eventuate. The deal – we don't know the details – might have been discussed at this time.

Macarthur's restless urge to trade continued, and in June he informed King that he had bought a brig and asked his permission to use it for trade in the Friendly Islands. He pointed out that an American ship had recently obtained a great deal of sandalwood from that source, and he was keen that the profits of the trade be rescued from the hands of foreigners and, here referring to Simeon Lord and other emancipist ship owners, men whose *loose and immoral characters threaten to produce the most fatal effects upon the rising generation.* King agreed that Macarthur's ship could go to Fiji to collect sandalwood, but it could not then sail to Canton, where there was a heavy demand for the wood. This was but one incident in the beginning of the efforts by Macarthur and others to circumvent the East India Company's monopoly of trade with Asia.

Edward Macarthur, now aged seventeen, returned to New South Wales in August to spend some time with his family before going back to Britain to commence his military career. On the 6th of the same month William Bligh arrived. The handover took six days, and during this period King granted his successor 1,000 acres on the Hawkesbury and two smaller lots of land at Parramatta and Petersham Hill. He also granted Mrs Mary Putland, Bligh's daughter, 600 acres, and backdated the deed by seven months. Bligh returned the favour by granting King's wife 790 acres. (The 2,350 acres King gave his children had first been given to his wife, until he found out this was not permitted.) All these grants were illegal. In general King was an honest man, but he was anxious about his poverty and the future of his family if he died soon, which, given his state of health, seemed possible.

It's easy to see such transactions as scandalous, but it must have been galling for the early governors, who were not particularly well paid and whose jobs were difficult and tenuous, to see men like Macarthur growing rich around them. King had been a competent governor, albeit more effective in his early years. He had restricted the number of convicts provided to the officers, reduced the number of people living off the government store and cut the public cost of running the colony significantly. He encouraged emancipists with good jobs and introduced an early version of the ticket-of-leave system, providing early pardons to convicts who behaved well. In 1804 he had sent William Paterson to successfully found the colony at Port Dalrymple. Although much worn down by the time he left, King had done a good job in difficult circumstances.

Government House, 1791
Due to the shortage of lime with which to make proper mortar, this for some years was the only two-storey building in the colony.

The early colonists were struck by the flora and fauna of New
These pictures by John Hunter were done during his first stay in the
creatures he was one of the first Europeans to draw: the

South Wales more forcefully than we can imagine. colony. They show him struggling to capture the shapes of kangaroo, emu, eastern rosella and kookaburra.

John Macarthur: soldier, pioneer, farmer, trader, amateur lawyer and architect, shipowner, family man, duellist, administrator, rebel, plutocrat, political thinker, founding father, man of honour.

Elizabeth Macarthur, who lost two children in infancy and brought up seven others. As with Mary Reibey and Caroline Chisholm, the absence of her husband (John was away for twelve years) relieved her of constant pregnancies and enabled her to do other things, in this case running the farms and looking after the family business interests.

Elizabeth Farm. This 1820s picture is the earliest
three main rooms and an entrance hall, and there was no
(which is open to the public) and is the

we have. When built in 1793 it contained only
verandah. A part of the original remains in the present house
oldest European building in Australia.

The Vinegar Hill convict uprising of March 1804. This unusual
No one was hanged from trees, although the leader Phillip Cunningham was
In the foreground, Cunningham cries Death or liberty Major

picture shows different stages of the fight occurring simultaneously.
hanged, without a trial, from the staircase of the storehouse at Green Hills.
and Johnston responds, You scoundrel, I'll liberate you.

Sydney in 1808, seen from what is now East Circular
Orphans House. Two days later, Major George Johnston
and marched them along Bridge Street (mostly

Quay. Macarthur's trial on 24 January occurred in the assembled his regiment to the left of St Phillip's Church not visible) to Government House, on the far left.

Sydney 1808, to the right of the previous picture, showing The Rocks and Dawes Point, where the southern end of the Harbour Bridge now rests.

Government House, Parramatta. Its much-admired garden was the scene of some interesting events. A dispute over the theft of turnips in 1796 started the great feud between Richard Atkins and John Macarthur. And here in 1806 Bligh accused Macarthur of lying to the government about his wool, and told him: You have got 5,000 acres of land in the finest situation in this country. But by God, you shan't keep it! *The building shown here was begun in 1799 and, considerably extended since, is open to the public today.*

Governor William Bligh discovered by soldiers on the night of the afterwards, by Sergeant-Major Thomas Whittle, and put on display to from fear but in the hope of being able to escape and rally his

rebellion of 26 January 1808. The picture may have been done shortly damage Bligh's reputation. In fact, the governor almost certainly hid not supporters. The attempt failed, to Bligh's everlasting detriment.

Government House, 1807. Not a large building – that it took the rebels up to two hours to find Bligh suggests he must have been hiding, despite his later denials.

London

Back in England, Sir Joseph Banks was working on Bligh's behalf, in particular searching out information on John Macarthur. Among his papers is a memo to himself, probably written in June or July 1806, concerning queries he was putting to Captain Kent. How had Macarthur purchased some of his sheep? *If the purchase of any one of them could be traced it would be advantageous.* As for the Cowpastures, *are they not more fitting for neat stock than for sheep*?

What did Macarthur do to cause his arrest by the governor?

Who was Macarthur's father and how was his youth employed?

The angry knight also sought information from Captain Henry Waterhouse. Once the replies were in, he prepared a snarling denunciation of the hated sheep-monger. The letter was written for *their Lordships*, although we don't know which ones, to be considered before any more land might be given to Macarthur. (Banks was probably concerned that he would ask for the second 5,000 acres originally offered by Camden.) The letter's subject was Macarthur's *pretensions to extraordinary indulgence from government on the plea of having raised a flock of fine-wooled sheep at Sydney*. Banks argued that Macarthur had lied about his own importance in sheep-breeding and that he was no more entitled to a large grant of land than were other colonists who had had equal success in breeding sheep. Indeed, none of them was entitled to such a grant, as *the success of their concerns will enable them to extend their concerns [presumably by the purchase of land] full as fast as their flocks can be expected to increase.*

Banks noted that Macarthur was the son of a mercer and tailor and that his recent *visit* to England, *as he modestly terms it, was under an arrest of the governor for his extreme ill-conduct in the colony.* It is an interesting reflection on information flows of the time that Banks was still, despite his deep interest in the matter, not aware of the details. He then accused Macarthur of making a huge fortune from spirits-trading, indeed of having obtained his first sheep in this manner *from individuals to whom they had been granted by government as a farming start as they had shown symptoms of a reformation*. Moreover, *some if not all of these unfortunate wretches who when Captain Macarthur thus obtained their little property at that time were beginning to lead new lives and reap the comforts of honest industry, were plunged back by despair arising from their imprudence into the abyss of vice and misconduct from whence they had so lately emerged*. Rising to a climax of denunciation, Banks speculated that *some of them are now perhaps working at the brick cart as degraded slaves who had Captain Macarthur never been born would have by this time have become useful and honest settlers.*

Banks suggested that in general *his conduct while resident in the colony does not merit reward. ... That a man of this character should have obtained leave to return on any terms to the governor whom he had insulted and to the colony where he had practiced the conduct above stated, is a clear proof that government acted under a misrepresentation of his conduct when they gave him any kind of indulgence.*

We can assume that William Bligh would have been sent a copy of this extraordinarily bitter attack, and that it would have helped shape his attitude to Macarthur thereafter.

VII

Bligh

William Bligh came from a line of minor Cornish gentry – his father had been a customs officer – and he was fifty-two years old. It was not the first time he had been on Australian soil. He had brushed Tasmania on his way to the Pacific and, in the launch after the mutiny on the *Bounty* in 1789, had actually considered heading for Port Jackson. However, being uncertain if the fledgling colony survived, he decided to head for the Dutch East Indies via Cape York instead.

Bligh was famously irascible, so it's remarkable that he had come this far without fighting, or even coming close to fighting, any duels. The main reason for this was that he had spent his adult life on ships, and naval officers fought fewer duels than military ones. Duels usually arose from situations where uncertainty existed, over words or actions. Often they were between men equal in status, even if questions about status were involved. But such circumstances were far less common at sea than on land. On a ship, a man's rank in the hierarchy was known to everyone, as were the duties expected of him in almost any situation that might arise. Behaviour was regulated and monitored – ships were crowded and it was a very public life, in which there were usually witnesses to arguments and actions. So for naval officers, the potential for the moral ambiguities on which the code of honour thrived was much reduced. This is why in the *Bounty* mutiny the issue of respect had been subsumed by the issue of obedience.

Unfortunately for Bligh, some of the *Bounty*'s crew had powerful relatives and friends, and while the subsequent trial in 1792 found against the mutineers, three of whom were hanged, he gained many enemies from the experience. The naval authorities had backed Bligh, then and later, because discipline had to be upheld among the lower orders. But this did not stop his enemies from trying to harm him, just as his patrons tried to help him. Alan Atkinson has shown that, quite apart from the *Bounty* affair, Bligh became enmeshed in a far-reaching feud involving the civil administration of the navy. The details of this need not concern us, but it meant that by 1806 Bligh had an unusual number of bitter enemies in and outside the navy, and the new foes he was to make in New South Wales were able to take advantage of this. Indeed, from the moment he was appointed governor there were powerful people keen to see him disgraced. At one point, even Sir Joseph Banks was *literally hooted out of the board room* when he approached the Admiralty while negotiating the details of Bligh's appointment.

Given Bligh's reputation as a hard man, his selection suggests that the British government continued to believe that while the New South Wales Corps was out of control, it could be reined in by a strong governor. The fact that the corps was, after fifteen years, so entrenched and powerful that it was beyond the capacity of any governor to bring it into line had not yet sunk home in London. They were throwing Bligh to the wolves, as they had previously thrown Hunter and King. Yet Bligh was made of different material. He was intelligent and healthy, worked fourteen-hour days, was abstemious in his habits and didn't mind making enemies.

He had sailed to Sydney in command of the *Lady Sinclair*, along with another ship, the *Porpoise*, commanded by Captain Joseph Short. There was confusion over which of the two captains had control of their little convoy at any one time, and they argued on and off after leaving England. At one point Short even fired a shot across Bligh's bow. Short intended to settle in New South Wales, but on arrival was accused by one of his officers of drunkenness and cruelty, and Bligh asked ex-Governor King, and Lieutenants John Houston and John Oxley of the *Buffalo* to investigate the matter. They found that Short had a case to answer, so Bligh sent him home to a court martial.

On his arrival in Sydney, Bligh had been greeted with an address of welcome signed by George Johnston for the military, John Macarthur for the free settlers, and Richard Atkins for the civil officers. Soon after this, Bligh received two petitions from a total of 369 settlers complaining that Macarthur was no representative of theirs, as he had recently withheld some of his sheep from market in order to drive up the price of mutton.

Historian Anne-Maree Whitaker has suggested that Bligh's knowledge of the colony was seriously out of date, and was based on ex-Governor Hunter's 1802 book about New South Wales. This could explain his anachronistic belief that the colony's main problems were alcoholism and the officers. It

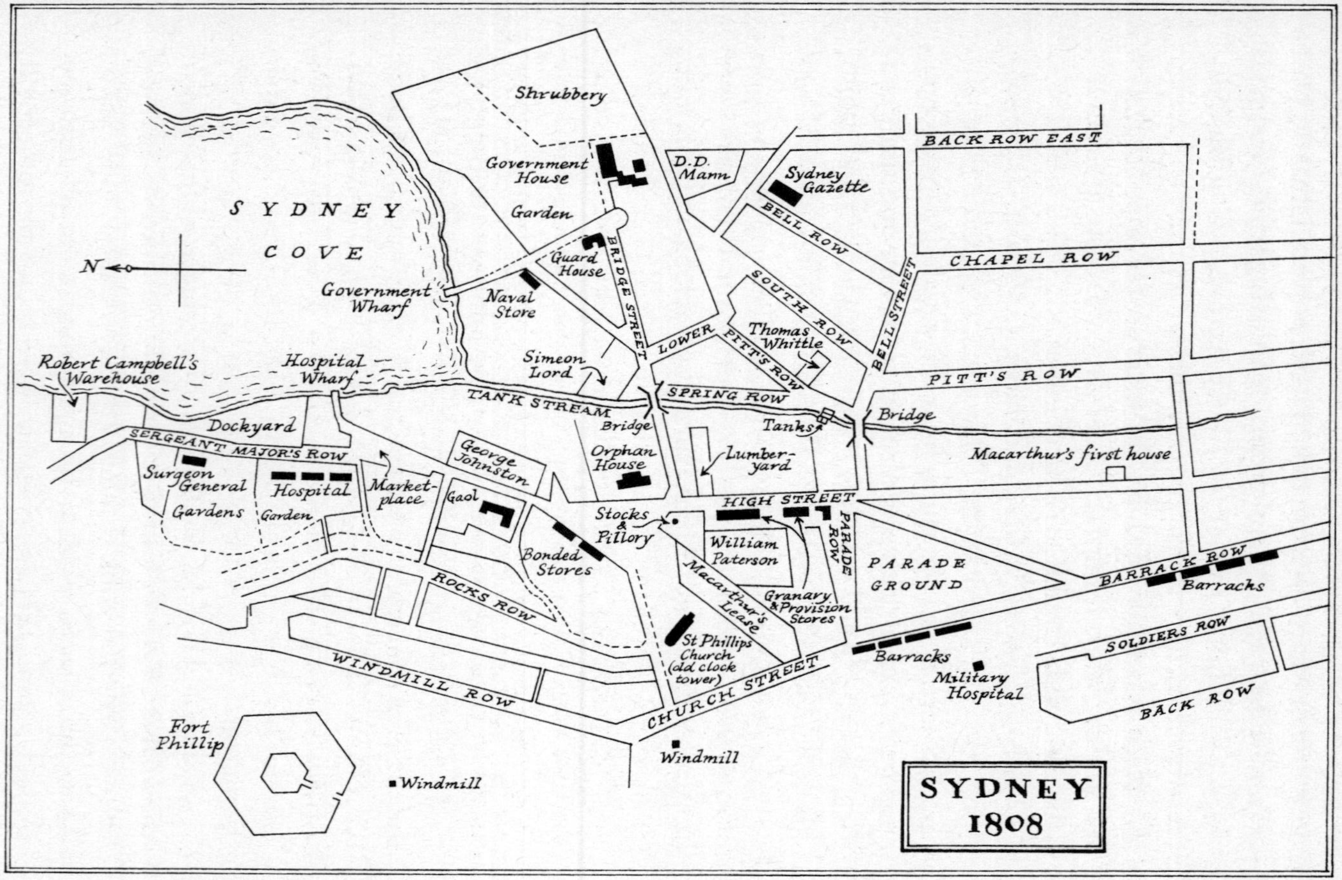
SYDNEY
1808
SYDNEY
COVE
N
Shrubbery
Government House
Garden
D.D. Mann
Sydney Gazette
BACK ROW EAST
BELL ROW
CHAPEL ROW
Guard House
Naval Store
BRIDGE STREET
Government Wharf
SOUTH ROW
BELL STREET
Thomas Whittle
LOWER PITT'S ROW
PITT'S ROW
Simeon Lord
Robert Campbell's Warehouse
Hospital Wharf
TANK STREAM
SPRING ROW
Bridge
Tanks
Bridge
Dockyard
SERGEANT MAJOR'S ROW
Orphan House
Lumber-yard
Macarthur's first house
George Johnston
Surgeon General
Hospital
Market-place
Gardens
Garden
Gaol
Stocks & Pillory
HIGH STREET
William Paterson
PARADE ROW
PARADE GROUND
Bonded Stores
Macarthur's Lease
Granary & Provision Stores
BARRACK ROW
Barracks
ROCKS ROW
St Phillips Church (old clock tower)
Barracks
SOLDIERS ROW
WINDMILL ROW
CHURCH STREET
Military Hospital
BACK ROW
Fort Phillip
Windmill
Windmill

meant that in some of the fights Bligh immediately picked with the military officers he was like a general fighting the last war instead of the current one. Another possible reason for his focus on the officers and alcohol might have been the influence of Banks' obsession with Macarthur.

Bligh moved to replace most of the colony's officials with his own appointments, most importantly dismissing the corps' surgeon, John Harris, from the position of naval officer, and appointing Robert Campbell in his place. Campbell was the colony's biggest trader and there was a conflict of interest in giving him control of the harbour, although there's nothing to suggest he abused the position. This didn't stop his competitors like Macarthur from resenting the situation. Bligh formed a close relationship with Commissary John Palmer, who was also a leading businessman, and whose sister Sophia was married to Campbell. Thus the new governor reduced the power of the military and established what must have seemed to them like a closed circle from which they felt excluded.

In October 1806 Macarthur's various holdings of 850 acres around Elizabeth Farm were consolidated by a new covering deed. His property had grown since the first grant of 100 acres in 1793 by means of subsequent grants and also purchases. One example of the trail of property dealings involved is as follows. A man named Stewart had been granted land by Governor Phillip in 1792 and had sold this to a former ensign named Cummings, who received two grants himself from Grose. Cummings also bought land from a soldier named Richardson and another man named Roberts, who had received their grants from Paterson. Later Cummings sold the whole lot to Macarthur, who also bought and sold other property. It can be seen from this that one reason Macarthur became a major landowner was simply through persistence. Others had started with similar opportunities but dropped out along the way.

In early 1807 Bligh banned the use of spirits in bartering. His instructions from the government noted that *great evils have arisen from the unrestrained importation of spirits* into New South Wales, and ordered him *on pain of our utmost displeasure* to reduce the amount imported and the trade that resulted. Bligh sought to prosecute this policy with vigour. The spirits that were permitted to be landed were distributed pretty equally among all the free settlers, so it was more difficult for the officers to corner the market and resell alcohol at an enormous mark-up. However, there might have been other routes open to them. Captain William Dorr of the American vessel *Jenny* later claimed that officers of the New South Wales Corps forced him to sell them spirits from his ship after Bligh had said they couldn't be landed. Possibly the story was invented as a revenge after Dorr was charged by Johnston with smuggling, but it went like this. Captain Kemp had interrupted Dorr at dinner one night and taken him for a drive in his chaise to George Johnston's house. There he waited while Kemp and the major took a walk in the garden. Then Kemp and he had returned to Sydney, and on the way *he informed me that he had been talking with the major respecting my spirits, and said that the major had given his permission, but wished that his name might not be brought in question.* Following this curious outing, Dorr had *been continually harrassed by those gentlemen officers, and men employed under them, to undertake the smuggling of spirits.* Lieutenant William Lawson was one of those concerned. Dorr said that he'd been told by a fellow captain, Corry of the *Eliza*, that John Macarthur and his trading partner Garnham Blaxcell had acted in a similar fashion towards him.

The trade in spirits had played an important part in commercial life – and the enrichment of John Macarthur – since 1792. In 1808 Dr Robert Townson, a free settler who was no enemy of Macarthur's, observed to Viscount Castlereagh that *There is not, my Lord, a single man in the colony who has made even a small fortune by farming or any kind of agricultural pursuit. These fortunes had their foundation laid in the earlier times of the colony, when the officers, civil and military, were allowed an underhand kind of traffic in spirits, &c., and were also allowed a great many men victualled at the expense of the Crown.* This seems an accurate description of wealth creation in the 1790s, although by 1808 the sale of meat was probably a superior source of income for the gentlemen farmers. But Macarthur continued to sell spirits. One deal in May 1807 involved the purchase of 270 gallons of spirits from the commissariat for seventeen shillings a gallon. After disposing of some of this, he mixed the rest with cherry and orange juice and water and sold it for forty shillings a gallon.

Meanwhile Bligh moved around the colony, working hard to impose his will. Like King he had a coach pulled by four horses and travelled with at least half a dozen soldiers as his bodyguard. He introduced another reform which affected the pockets of the traders. He ordered that private promissory notes or notes-of-hand (effectively IOUs), which had previously been redeemable according to the value of certain goods, could now be made out only in their sterling value. Bligh also moved to stop the flogging and imprisonment of convicts without a magistrate's warrant. According to Provost Marshal William Gore, this had been prevalent under Governor King, as had the practice of flogging convicts to extract confessions. Altogether, the above reforms served to irritate the colony's wealthy inhabitants and mollify the poor.

They came with an unfortunate style. Bligh was described by a later judge advocate, Ellis Bent, as *a broad, stout little man ... more violent of temper than I could have conceived.* According to Bent, Bligh's daughter, Mary Putland, had a *very small, nice little figure* but she too was *extremely violent and passionate, so much so now and then as to fling a plate or candlestick at her parent's head*. We can see why Mrs Bligh might have decided to stay in England.

While some of the poor would have been happy with the restrictions placed on the activities of the establishment, they received little enough from Bligh themselves. The governor was particularly mean with the greatest gift in his power, land grants. Bligh was a man loath to let go of what he had. It would have been better for everyone had he paid more attention to the development of the colony, and less to largely imaginary fears of rampant alcoholism and trading monopolies. As it was, he suppressed the outlets for the colonists' energy, which raised the general level of resentment and tension.

Bligh was so prone to anger and violent language that a man's first approach to him on some delicate subject would often evoke a response of such rage and fury that you'd think the governor had been importuned about it for months and had been driven beyond endurance. Macarthur too, of course, was given to anger and enmity, but in him it was deeper and generally restricted to competitors and people who threatened his interests or his sense of honour. With Bligh, anger was a general policy, and he directed it at many people, often for no reason at all. As noted earlier, with him everything became personal; he had the ability to make men feel like children, and they resented him for it. This gradually poisoned his relationships with most of the powerful men of the colony, as throughout 1807 he offended them one by one, both in the way he talked to them and what he said.

He started early, with Macarthur. In September 1806 Bligh was visiting King and his family, who were staying at Government House at Parramatta until their ship was ready to leave.

Macarthur decided to call on the governor to discuss his sheep-breeding activities. According to Macarthur's account, he found him walking alone in the garden and tried to engage him in conversation, but Bligh angrily retorted, *What have I to do with your sheep, sir? What have I to do with your cattle? Are you to have such flocks of sheep and such herds of cattle as no man ever heard of before? No, sir!* Macarthur replied that he'd been told the British government had recommended him to Bligh's notice, but the governor expostulated again: *I have heard of your concerns, sir. You have got 5,000 acres of land in the finest situation in this country. But by God, you shan't keep it!*

Macarthur pointed out he'd been given the land on the recommendation of the Privy Council (not true) and by order of the Secretary of State.

Damn the Privy Council, replied Bligh, *and damn the Secretary of State too! You have made a number of false representations respecting your wool, by which you have obtained this land.* Apparently Bligh had been well briefed by Banks; even so, Macarthur denied the charge and asked the governor to come over to Elizabeth Farm that morning to inspect his sheep for himself. The two men entered the building at this point and sat down to breakfast with Mr and Mrs King. Bligh, according to Macarthur, *then renewed the conversation about my sheep, addressing himself to [ex-] Governor King; when he used such violent and insulting language to him, that Governor King burst into tears.*

After breakfast the three men and Major Edward Abbott visited Elizabeth Farm and inspected the flock of sheep there. Bligh then asked what the purpose of the examination might be, as nobody had ever doubted the possibility of raising fine wool in New South Wales. Macarthur replied that he thought Bligh himself had done so that morning. *No such thing!* Bligh responded passionately. *And I desire, sir, you will never attempt to attach any such meanings to my words.* Macarthur again raised the matter of the Secretary of State's support for his undertaking, and Bligh said *Damn the Secretary of State, what do I care for him, he commands in England and I command here.* The party moved indoors. Bligh paid his compliments to Elizabeth Macarthur and departed, leaving the others to express *the greatest surprise and astonishment at the violence they had witnessed.*

The above story was told by Macarthur in London at George Johnston's trial for mutiny in 1811. Some of it might be wrong – a number of the witnesses there lied through their teeth. Yet deciding where the truth lies is not easy. Bligh was a genuinely choleric man, and it was a time when implausible events did occur. Some of the account seems artful, but other parts seem so absurd it's difficult to see why Macarthur would have made them up, especially as there were people following the trial (e.g. King's widow) who might have exposed any lies. In any event, the great falling-out had begun. Four months later Elizabeth Macarthur was telling a friend in England that Bligh *has already shown the inhabitants of Sydney that he is violent, rash, tyrannical. No very pleasing prospect at the beginning of his reign.*

Bligh's enthusiasm remained fixed on the small settlers, who were desperate enough to be grateful for anything he did for them, and poor enough never to threaten his sense of authority. In December he announced that anyone – meaning any landless labourer – who refused to work at getting in the harvest would be set in the stocks for two days and one night. A further refusal would lead to three months' hard labour. He also fixed the rate for such work, and threatened with punishment any employer who dared pay more. This measure was due to the *evil tendency* of some workers to demand higher wages than the farmers could afford. For good measure, any worker asking for more would also receive the above penalties.

1807

According to Charles Walker, who commanded a brig belonging to Macarthur, in early 1807 Bligh refused Macarthur some item from the government stores, and Macarthur complained that the governor was giving stores to ordinary settlers, *a set of rascals who would deceive him; it would be better if he gave it to me and some of the other respectable gentlemen of the colony; if he does not, he will perhaps get another voyage in his launch again*. In February Bligh wrote to William Windham, Secretary of State for War and the Colonies, renewing King's efforts to have Macarthur's 5,000 acres grant moved from the Cowpastures. We don't know if Macarthur was aware of Bligh's letter to Windham, but if he was (or became so later) it would not have increased his regard for the governor. Given that large land holdings were so important, both to wealth and social status, and that the Cowpastures was the best land so far discovered, it is quite possible that this opposition contributed to Macarthur's subsequent actions.

We have seen how Macarthur sought to live the life of a gentleman in a penal colony by sending his boys to England for their schooling. In Australia he provided them with tutors. So in the period 1807–1809, a French Catholic royalist and former member of the New South Wales Corps, Gabriel Louis Maries Huon de Kerrilleau, taught James and William at Elizabeth Farm. This early Australian intellectual enjoyed reading while walking, which presented certain navigational problems. Later in his life he embarked on a long trip from Corrundaroo to Campbelltown, head down in a book. He was seen veering into the bush and *twice put on the right road, but apparently became lost in gullies of the Shoalhaven River: he was never seen again*.

Macarthur and his partner Garnham Blaxcell had interests in a number of ships around this time, which were involved in an array of trading ventures. In March 1807 the *Elizabeth* went to Fiji as part of a multi-ship venture involving sandalwood. The plans were upset by storms and in November she returned to Sydney with pork from Tahiti. The partners also owned parts of the *Halcyon*, *Parramatta* and *Dart*, a sealing vessel. What little we know of Macarthur's many business ventures suggests a man of unusual vigour and ingenuity, always looking for something new, always with several schemes on the go.

In April John Blaxland arrived in the colony. He and his brother Gregory, who had come out the year before, were the first free settlers of substance to emigrate to New South Wales, and did so after being promised certain benefits by the government. Most importantly for John was the guarantee of 8,000 acres of free land. In return for these benefits, he promised to commit £6,000 of capital to the colony. But he received only 1,290 acres of land from Bligh, and only twenty-three of the eighty convicts he had been promised. Bligh's reason for not providing the rest of the acres was that Blaxland had not committed the requisite amount of money to agriculture (which the government was keen to encourage), but had spent most of it on other activities, opening a butcher's shop in the town and running the *Brothers*, a whaling vessel. The Blaxlands even approached Bligh asking for permission to establish a distillery and offered any of Bligh's friends a cut in the company they were proposing to form. Bligh professed to disapprove of this approach, although several of his own supporters, such as John Palmer, were sharp businessmen and used their connection with the governor to their own advantage. Maybe the Blaxlands were just hoping for a slice of the action.

A general problem in the wealthy settlers' dealings with Bligh was the shortage of convicts. Only 189 male convicts came out during Bligh's time, the Napoleonic Wars' need for manpower still diverting most criminals into the armed services, so he was literally unable to fulfil the promises made by the authorities in London. Of the first three convicts Blaxland received, *one was an aged man, with one eye and one arm; another was an aged man with an asthma, who appeared to be about 60, and incapable of work; the third was an idiot.* Blaxland was also to have had a certain number of cattle from the government herd, but here too Bligh did not provide what he had hoped for. So the Blaxlands turned against him.

Macarthur's open breach with Bligh occurred in July following a legal dispute with Andrew Thompson, the manager of Bligh's farm at the Hawkesbury, over a promissory note. The value of the note had been expressed as a certain quantity of wheat. The value of wheat then rose dramatically after the Hawkesbury floods. Thompson wanted to pay the cash value of the wheat at the time the note was made out, but Macarthur insisted that he pay its current value. The case went to court and Macarthur lost. Alan Atkinson has noted that Macarthur took this action soon after the arrival of the *Young William*, the first ship to arrive with news that Lord Mulgrave had become First Lord of the Admiralty. Mulgrave was a friend of General Francis Grose, Macarthur's old friend, the former head of the New South Wales Corps. Perhaps Macarthur felt this freed him to begin his attacks on Bligh, and picked the dispute as an indirect way of getting at the governor.

Until then, despite John's differences with Bligh, the Macarthurs had frequently entertained him, his daughter and his son-in-law at Elizabeth Farm, and Macarthur had often dined at Government House. Bligh even said he wished he had a spare bed to accommodate Macarthur when he was down in Sydney. But after Bligh refused to hear Macarthur's appeal following his clash with Thompson, Macarthur stopped going to Government House. His wife did not, and one day at Parramatta she was asked where her husband was. Elizabeth, perhaps a bit flustered, replied that he was unwell. Bligh called in at Elizabeth Farm the next day and found John coming in from a ride on his horse. A certain amount of embarrassment followed.

What had particularly upset Macarthur was an article inserted in the *Sydney Gazette*, presumably at Bligh's command. Without referring directly to the Macarthur–Thompson case it described the general situation involved in the promissory note case as one *when Shylock still insists on his bond*. Two letters appeared in subsequent issues of the newspaper from someone, presumably Macarthur, signing himself 'An Occultist'. He suggested the editor abstain as much as possible from calling names, specifically *the opprobrious epithet of 'Shylock'*.

At the same time that Macarthur fell out with Bligh, D'Arcy Wentworth was found guilty of refusing to treat two sick convicts at the Parramatta Hospital and was publicly reprimanded by Major Johnston before the troops. Bligh then dismissed Wentworth from his position as assistant surgeon, without any explanation. There was a complex background to this action, and Wentworth was by no means innocent, but some observers chose to interpret this as Bligh unfairly adding a new punishment after the court had made its decision.

In 1807 there occurred the trial in Sydney of Michael Dwyer and others for planning an insurrection. Dwyer was an Irish gentleman rebel who had surrendered to the English on the agreement he be exiled but not charged. Following the result of the trial, where Dwyer and four others were found not guilty, Bligh hauled them before a bench of magistrates who did find them guilty and sentenced them to transportation. This action upset a great many people in the colony, including George Johnston, and made them ready to assume the worst of Bligh where his use of the legal system was concerned. In Bligh's defence it might be ventured that he thought the men were *prisoners for life*, so he could send them wherever he liked. That, at least, is how he portrayed matters in his despatch to London. But by now the governor was in the position where he was so unpopular that everything he did was interpreted by many in the worst possible light. He became gradually aware of this, but his sense of self-importance prevented him from moving beyond it and attempting to regain the confidence of those he had alienated. He persisted in believing that being right and being effective were – or ought to be – the same thing.

We have seen that Governor King gave out leases of land previously set aside for the Crown by Phillip. Some of it was in the Domain by Government House. Bligh was keen to improve the grounds of that residence; Mrs King had laid them out in a formal, geometric pattern but he was an enthusiast for the currently fashionable Picturesque style. Needing space, he ordered the leaseholders to pull their houses down. He offered them alternative land, but stories vary as to whether they were to be given new leases, and whether they would receive compensation for the loss of their houses. One of those involved was David Dickinson Mann, who said later that when he told Bligh he could not afford to abandon his £400 house, and that by the laws of England he could not be forced to do so, the governor yelled: *Damn your laws of England! Don't talk to me of your laws of England. I will make laws for this colony, and every wretch of you, son of a bitch, shall be governed by them. Or there [pointing towards the gaol] is your habitation!* Numerous witnesses later confirmed Mann's claim that this and similar incidents spread great concern throughout the colony regarding the sanctity of property. Sergeant-Major Thomas Whittle claimed the governor had visited him in Spring Row one morning at 7.00am and told him to pull down his house by 10.00am. *Sir,* Whittle said, *I have got a lease of this house.* To which Bligh replied: *I will have the house down again by ten o'clock, and you shall neither take bricks nor any thing else away.* The canny Whittle signed his lease over to Major Johnston, and the house stood, the incident presumably giving the commander food for thought.

Bligh was concerned to clear another area of Sydney, around St Phillip's Church. In terms of planning, this desire was entirely warranted, and would probably have been accepted at a later period more readily than it was in his time. The need for the common good to override personal interest is fairly widely acknowledged today, in principle if not always in practice. But in 1808 the relationship between government and property had entirely different connotations. Basically, property was considered the cornerstone of liberty. So when Bligh tried to turn maybe ten people out of their homes, many other inhabitants of Sydney were genuinely disturbed. Even if Bligh was in the right according to the strict letter of the law, his actions were seen as offensive to its spirit, and as threatening considerable injustice. People began to mutter. Abbott later recalled that Bligh's behaviour around this time was *very oppressive and ungentlemanlike ... I never came down from Parramatta, but what there was some complaint or other against him.* Richard Atkins recalled that *the people complained very much, whenever I went along the streets, the people were talking about those houses being pulled down* and wondering who would be next.

In August 1807 it was the turn of the successful emancipist traders to be alienated. Simeon Lord, Henry Kable and James Underwood wrote to Bligh requesting permission to transfer some goods directly from one ship to another without landing them first, which was illegal. Bligh responded by bringing them before his tame magistrates' court, where they were each fined £100 and gaoled for a month, merely for making the request. It seems an extraordinary punishment.

The general situation continued to deteriorate. There was an attempt to hurt Bligh by fabricating two charges against Provost Marshal William Gore, effectively the colony's chief constable and gaoler, responsible for enforcing the regulations against spirits trading. The two petty charges went to court, where they were thrown out for lack of evidence. One of the witnesses was the wife of a gaoler Gore had threatened to sack for brutality. She did not turn up at the trial, and Surgeon Thomas Jamison swore she was giving birth. This was apparently wrong, and Bligh later dismissed Jamison from the position of magistrate. This attack on Gore, like the similar attack on Isaac Nichols some years before, illustrates the lengths to which the officers were prepared to go to punish those who sided with the governors.

To fight back against the forces arrayed against him, Gore used the services of George Crossley, a former London lawyer transported for perjury. His foes were outraged by this reliance on a former criminal. A gentleman's word was his bond: telling a lie was one of the worst things he could do. So it struck Macarthur and others as particularly repulsive that Crossley, being no gentleman, should be able to practise the law again. Had he gone into trade they would have been less concerned, but the law was sacred. It would have made little difference were Crossley a reformed character, but in fact he wasn't. Having arrived in Sydney in 1799, the following year he had drawn dishonoured bills of exchange, defrauding one person of £2,000.

In September 1807 Captain Anthony Fenn Kemp arrived in Sydney from Van Diemen's Land and found the colony in *a very disturbed state ... every person I associated or conversed with complained to me of the tyrannical and arbitrary conduct of Governor Bligh. Messrs. Blaxland and Captain Bunker, formerly in the service of Mr Champion, as master of a South Sea whaler, told me, they had embarked their all to come to New South Wales, under the assurance of Government, and with letters they brought from Mr Cooke, the Under Secretary of State, to Governor Bligh, that they were to have specific encouragement of land and stock.* This they claimed not to have received, and as a result they considered themselves and their families ruined. Other people, including Garnham Blaxcell and Nicholas Bayly, D'Arcy Wentworth, and Dr Robert Townson, told Kemp *that Governor Bligh threatened to pull down houses in every direction, and to take leases away; therefore they considered that no man's property was safe under such a government.* Major Edward Abbott, by no means an uncritical supporter of the subsequent rebellion, said that several houses were pulled down at this time by Bligh's order, and many people feared to build on their leases or extend existing buildings: *every person who had property felt an unpleasant sensation.*

While the horrors of the French Revolution were of course fresh in everyone's minds, when educated Britons in the early nineteenth century spoke of freedom, oppression and rebellion, they were thinking not so much of what these words mean to us – mass rebellion, complete democracy and almost total freedom – but the conflict between a monarch and the upper classes. This conflict was focused on property rights and the laws that protected them, for property is the basis of democracy. (Excessive taxation, incidentally, was regarded as an interference with such rights.) In 1690, two years after the Glorious Revolution, when the British replaced James II with William and Mary, the very influential philosopher John Locke had written that a man submits to government by choice, to protect his life, liberty and property. Locke also established the idea that rulers have responsibilities. *Whoever has the legislative or supreme power of any commonwealth is bound to govern by established standing laws, promulgated and known to the people, and not by extemporary decrees, by indifferent [impartial] and upright judges who are to decide controversies by those laws*, he wrote. Bligh trampled on the conventions of the colony regarding property, and gained a reputation for relying on a judge advocate and magistrates who were too ready to do as he wanted.

This link between property and freedom in the minds of Macarthur and many of his contemporaries cannot be underestimated. One of the causes of the American Revolution was George III's refusal to allow the colonists to settle beyond the Appalachians. Although the Declaration of Independence speaks of *life, liberty and the pursuit of happiness*, the rebels' original call, in 1775, was for *life, liberty and property*. The Constitution of Virginia, devised the following year, claimed that all men have the right to *the enjoyment of life and liberty, with the means of acquiring and possessing property*.

This helps explain why Bligh's arguably reasonable town-planning action, in requesting that fewer than a dozen houses be moved, created such indignation, and was brought up against him again and again over the next few years. George Johnston would later assert that *property is too sacred to be taken away … merely at the will of a governor*. This is not to say those who used this as a casus belli for rebellion were necessarily right in doing so, although they might have been, but at the least they knew it would catch people's attention, and provide their cause with some ready sympathy.

In October Surgeon John Harris wrote to Philip Gidley King in England about the state of the colony. Dismissed from the posts of naval officer and magistrate, he was not happy.

Bligh had reduced the quantity of spirits imported so much that *He is determined that I shall be sober, for I have not had a glass of grog in my house for some months.* He had also *executed more men by three in his short time than you did in the whole of the time you was with us.* As for Bligh's crackdown on the wealthy citizens, *Had you been so severe the half of the long-coated gentlemen in the colony would have been in constant durance.* The pipe-makers were at work, and Harris recorded one effort: *Oh tempora! Oh mores! Is there no CHRISTIAN in New South Wales to put a stop to the tyranny of the Governor?* According to Harris, Bligh favoured his supporters, particularly merchant Robert Campbell, some of whom abused their positions. In one case *a man who lives in one of my houses must quit it immediately by order of the governor, and that no person should in future reside in it without his order. On enquiry I found that the cause of this order was that this man had some seed potatoes for sale which he would not let [Provost Marshal] Gore have without the cash … It is strange that the Governor will thus meddle with private property, which ever was and ought to be sacred.* Harris also wrote to Mrs King, telling her how at Government House *The shrubbery has also undergone a thorough change – no grass now growing in it, all laid out in walks with clumps of trees. … All the rocks in the garden is blown up and carried away. Not less than 80 or 90 men have been constantly employed since you went away for these purposes.* (If true, this would help explain why Macarthur and John Blaxland were resentful at not receiving more convict labour.) *Carriage roads are now all round Bennelong's Point, and down about Farm Cove, all ditch'd in and no thoroughfare allowed.*

On 8 October 1807 George Johnston wrote to the Duke of York, commander of the British army, complaining about Bligh's treatment of the corps. Again it was a matter of Bligh's style as much as what he actually did. The governor had abolished the roster system for appointing officers to court work and now made his own choices, and he had taken soldiers away for other duties, all without advising Johnston. He had criticised the corps to guests at Government House and insulted individual soldiers to their faces. Private Richard Mason, a member of the governor's bodyguard, had asked for another assignment because he'd had enough of Bligh calling them *tremendous buggers, wretches, villains* and other epithets. Bligh was interfering with Johnston's own independence and authority, and could not be reasoned with. *Governor Bligh*, he noted, *seems ignorant of any instructions or rules, but such as are dictated by the violent passion of the moment.* This was not how a gentleman behaved.

So by this stage, just over a year after arriving, Bligh had managed to alienate the entire corps by treating them with contempt. He despised its officers along with at least some of the rest of the colony's establishment, later writing to the Honourable Charles Greville in England: *You can have no idea of the class of persons here who consider themselves gentlemen.* Indeed, three weeks after Johnston's letter, Bligh wrote to London suggesting the New South Wales Corps be withdrawn. He pointed out that the soldiers had become dangerously intertwined with the general population, including the convict women, in fact they had become *a dangerous militia.* (In Sydney the corps and its dependants and close business associates comprised more than half the free residents.) His daughter wrote at the same time that Bligh was *dreadfully harassed by business and the troublesome set of people he has to deal with … there are a few we suspect wish to oppose him, as yet they have done nothing openly, though it is known their tools have been at work for some time, that is they are trying to find something in papa's conduct to write home upon … Mr Macarthur is one of the party, the others are the military officers.* However, the two groups continued to mix; just over a week after this letter was written, the

Macarthurs held what the *Sydney Gazette* called *a splendid entertainment* for a large party of officers and ladies, including the governor and his daughter.

Also in this epistolary month, Bligh wrote a letter to Sir Joseph Banks that shows he understood much that was wrong with the colony. *The most material thing to be done is to make every one confident he will enjoy a just and upright government – remove without delay the very unfit and very disgraceful Judge Advocate – remove without delay the N.S.W. corps and send them to India. Whatever soldiers are to be here, let them be soldiers, and not those who are without exception ingrafted with convicts. … The officers are so connected by property and intercourse with the emancipated convicts, both of men and women, their influence affects public justice. The Judge Advocate is a disgrace to human jurisprudence. The criminal and civil courts should be changed and regulated by rules approximating to those of England. – The officers of the Crown should be honourable men and at least we should have an Attorney General as well as a Judge besides some respectable lawyer as a solicitor. Let this be done and all will be well.* These recommendations were good, but impossibly expensive and troublesome. Apart from anything else, they would have required politicians to think of New South Wales as something more than a prison, yet for them that was still all it was. No government would act on Bligh's expensive request unless something happened to force its hand. And in the meantime, Bligh had no intention of compromising with the colonists, despite the fact the limitations of the court system affected all of them.

In this month Macarthur was sent two stills by his London agent, in breach of Bligh's regulations. To summarise what became a complicated affair, Robert Campbell, as naval officer, ordered that those parts of them in Macarthur's possession be seized; Macarthur refused to hand them over, so they were taken from him; and Macarthur took Campbell's representative to court for wrongful seizure of his property. The case is of interest because it was brought on quite unnecessarily by Macarthur, who used it as an opportunity to make an inflammatory attack on Bligh in the packed courtroom, claiming that *It would therefore appear that a British subject, living in a British settlement, in which the British laws are established by the Royal Patent, has had his property wrested away from him … It is therefore for you, gentlemen, to determine whether this be the tenor on which Englishmen hold their property in New South Wales?*

H.V. Evatt noted that as far as the law is concerned, *all this was rhetorical nonsense.* He also pointed out the important role of the courts in the *political* life of New South Wales. Far from being simply courts as we know them today, they were de facto and grossly imperfect legislatures and theatres, places for opponents to slug it out over trivial issues which, as in this case, could have huge symbolic importance. This system was messy and unsatisfactory, and made more so by the problems of Judge Advocate Atkins, the military make-up of the bench, and the ready availability of a stream of corrupt and untruthful witnesses. One thing it was not was an effective mechanism for resolving disputes.

Macarthur took full advantage of all these failings. He used the courts to try to destroy men who threatened his interests, and to elevate personal disputes into political ones, thereby gaining supporters. Evatt observed that *the courts were the true forum of the little colony. They had no competitors as a means of expressing individual or public grievances. There was no legislature, no municipal government, no avowed political association or party, no theatre and no independent press.* This is one reason why Macarthur's story, particularly the next part of it, involves so many court cases. The courts of New South Wales in the early nineteenth century were not the objective and pretty effective machines that we have today, where on the whole everyone agrees with the cases that go to trial and the results that are achieved. In the early 1800s Sydney's judicial system was a confused mess open to abuse, and Macarthur was to abuse it more than anyone else. His purposes were personal and political, rarely judicial.

Evidence of the failure of the courts, in our terms, lies in the fact that Macarthur actually won the case of the stills: the majority on the bench, including Atkins himself, found that Campbell's representative had no right to remove them.

Atkins was abused by Bligh for the result of this case. Edmund Griffin, Bligh's twenty-year-old secretary, had attended the court and reported to the governor that Macarthur's speech had been *inflammatory*. Bligh then ordered Griffin to report this to George Crossley, who was at first thought to be at the house of Nicholas Divine, superintendent of convicts, but was later tracked down at John Palmer's residence in Woolloomooloo. These men were all enthusiastic supporters of Bligh. At Palmer's place, an indictment against Macarthur was drawn up by Crossley.

Bligh was clearly out to get Macarthur. Griffin later said that Bligh opposed large land grants on principle, and had said it was *extraordinary that a gentleman of Mr Macarthur's property could not live quiet*. Bligh thought Macarthur's general behaviour *very improper* and had spoken frequently of how he'd like to expel him from the colony. According to Griffin, it was believed in the governor's circle that Macarthur wanted to be *the Oliver Cromwell of this country*, and it was hoped that if he could be charged with inflammatory behaviour (for which the punishment, were he found guilty, would have been a fine and imprisonment) *he might make his escape from the colony and not stand trial*. This is important because it shows Bligh was determined to eject Macarthur from New South Wales. Presumably

Macarthur was aware of the depth of the governor's antipathy by now.

In November 1807 Anthony Kemp was told by Simeon Lord, D'Arcy Wentworth and others, that a group of wealthy settlers had agreed to provide £1,500 for Macarthur to go to London to complain about Bligh to the Secretary of State. So, many of the colony's strongest men were now opposed to Bligh and discussing this with each other, and Macarthur was considered their leader. Kemp told Bligh about this, and the governor replied that they might be damned, for he was *as irremovable as Mount Ararat*. That same month, Charles Walker returned from a voyage in Macarthur's brig and was told by his master *that the colony had suffered Governor Bligh to reign long enough, there would soon be an alteration*. The court case over the stills seems to have drawn together the different discontents of various groups – house-owners, land-seekers, the military – thanks to Macarthur's use of it for political speech-making. There was a growing sense that all grievances were linked, which produced an air of tension, although not yet of crisis. The energies of the more creative colonists were being frustrated by Bligh's reforms, his economies and his meanness, and still he refused to make land grants. King had granted 60,000 acres, whereas Bligh gave out only 4,180, of which 2,000 had been to himself and his daughter.

Meanwhile, life went on around the Macarthurs' growing agricultural holdings. The threat to their property and livestock from escaped convicts, known by then as *bushrangers*, was so great that they employed guards and put the sheep in enclosures after sunset. In mid-November an accident occurred on some of their land near Prospect. John Griffiths and Simeon Donnally were guarding a paddock one dark night when they thought they detected bushrangers nearby. Leaving the elderly Donnally at his post, Griffiths crept around the fence, and eventually saw a figure among the sheep in the paddock. He called out but, receiving no reply, opened fire and hit the man. It turned out to be Donnally, who was deaf, and who had inexplicably left his post. He did not recover from his wound.

In the same week, Aborigines told three of the Macarthurs' stockmen that bushrangers were camping at Cabramatta Creek. The three rushed the camp and caught five bushrangers. Then, deciding to tie them up in order to get them back to town, they sent one of their number to fetch strips of bark to use as rope. Once he was gone the villains attacked the other two, who, in the words of the *Sydney Gazette*, *maintained an obstinate struggle until the other returned to their aid with his musket and a tomahawk; which he plied with such dexterity that three [of the bushrangers] were obliged to fly, one of whom was severely wounded by a blow on the head with the butt of the musket leaving the other two in custody, much wounded also.*

In December the *Parramatta*, a schooner of which Macarthur was part-owner, returned to Sydney after a voyage to Tahiti to buy salted pork. The owners of vessels working out of Sydney were required to pay a fine if any convict escaped on board. It emerged that, despite Macarthur's vessel being searched before departure by the provost marshal's men, a convict named John Hoare had stowed away and escaped during the voyage. The naval officer placed two constables in charge of the ship, which was not to be returned to its owners until a forfeit of £900 was paid.

Macarthur had been involved in a somewhat similar incident, probably the year before, when a vessel for which he was responsible took three convicts back to England. This, at least, is what Bligh wrote to Windham in October 1807. That event may well have strengthened the governor's determination to do something about the Hoare affair. Macarthur now claimed the fine was not legal, due to a technicality, and announced that he was surrendering the vessel to the government and would claim compensation from his insurers. This incident too he portrayed as an attack on private property, arguing that the naval officer had effectively deprived him of his ship. He then informed the crew that they would not be paid or fed by him from that day on.

On 14 December Macarthur received a letter from Judge Advocate Atkins requiring his attendance at Sydney to explain why the crew had come on shore in violation of the regulations, claiming that Macarthur had deprived them of their food. He ignored the request and fired off a reply to Atkins saying that he could not incur expenses for a vessel over which he had no control. In response, Atkins issued a warrant ordering Francis Oakes, the Chief Constable at Parramatta, to bring Macarthur before him on 16 December. The game was warming up.

On 15 December Oakes arrived at Elizabeth Farm with the warrant. For reasons unknown, he chose to deliver it at 11.00pm. Macarthur was called by one of his servants and met Oakes in the hall. He took him into a room where there were candles and, after giving Oakes a chair and a glass of grog, read the warrant. And so he found himself, *nearly at the hour of midnight, in danger of being torn from my family, and of being confined in a loathsome gaol, amidst all the most abandoned felons which the colony contained*. Hannibal Macarthur was present with young Edward, who was sent to fetch Nicholas Bayly, his father's friend. According to Oakes, Macarthur *made use of a great deal of improper language* and said, *Mr Oakes, you will acquaint the persons who have sent you here with the warrant … that I will never submit unto the horrid tyranny that is attempted, until I am forced; but I consider it with scorn and contempt, as I do the persons who have directed it to be executed*. This message was written down and given to Oakes. As he left, he heard Macarthur say, *Let them alone, they will soon make a rope to hang themselves.* This seems to have been a comment to the other people in the room, referring to Bligh and his supporters.

On his way home, Oakes was overtaken by Edward Macarthur who said his father would go to Sydney after all, and asked for the letter. Oakes refused to hand it over, and went on his way. The next morning he was approached by Hannibal Macarthur, who told him that last night Edward had not been acting on his father's orders at all, but in response to Bayly's suggestion as to how they might rescue Macarthur from the mess he was plunging into. When Oakes implied at a later date that Edward *had* been acting on his father's instructions, Macarthur accused him of trying to make him appear *pusilanimous*, something no man of honour could tolerate. The game was on and Macarthur, far from wanting to placate anyone, was now determined to force others to take sides and bring matters to a head.

Oakes proceeded into Sydney and delivered the reply to Atkins and Bligh, and later to the bench of magistrates, consisting of Johnston, Abbott, Atkins, and Palmer. Johnston mused out loud that Macarthur was a turbulent character whose actions were not always in his own best interests, and the bench ordered that Macarthur be arrested and placed in gaol. It's interesting, in light of subsequent events, to note that in this decision Johnston demonstrated no undue fondness for Macarthur, even if the order for him to be gaoled was perhaps more of a threat than a serious intention.

When he was seized in Sydney later that day, Macarthur insisted on being taken before Atkins, where he *behaved in a very improper manner*, according to Oakes, but was nonetheless given bail. On 17 December Macarthur appeared before the bench and was committed to face trial regarding the

Parramatta affair at the next sitting of the criminal court in January.

Four days later, Macarthur pulled a surprise out of his hat. He had obtained a bill for £26 6s that Atkins had made out fourteen years earlier. It had been dishonoured when presented then. (This was the same bill of exchange Macarthur had referred to in his first major attack on Atkins back in 1796; he must have obtained it then and kept it all this time.) He now requested payment of the original amount plus interest, to a total of £82 9s 5d. Atkins said he would pay but then disputed the amount of interest. A week later, Macarthur wrote to Bligh triumphantly requesting he appoint another judge advocate, as Atkins now had a conflict of interest.

By the end of 1807, then, Bligh had offended most of the colony's establishment deeply, essentially by doing things differently from how they were done in Britain. Although by nature an autocrat, he had actually tried to govern in a more objective and modern manner than was normal at the time. E.P. Thompson has argued that the eighteenth century was *a predatory phase of agrarian and commercial capitalism and the State itself was among the prime objects of prey*, and this situation continued into the early nineteenth century, at least. By interfering with established trading arrangements and, in particular, by opposing and even refusing to make land grants, Bligh was denying his more powerful subjects the prey to which they believed themselves entitled. If Bligh considered himself a modern ruler, objective and just, they saw him as a throwback to the time of autocracy. He couldn't see this because he had no experience of getting on with equals, of governing as distinct from giving and taking orders. Bligh probably had no idea of how provocative his actions and attitudes were. A mayor from a provincial town in England could have told him, but his advisers – merchants like Campbell, officials like

Palmer, the lawyer Crossley – were as ignorant as himself of the need to take strong people's interests into account if one wished to govern them successfully.

1808

Nothing occurred over Christmas, but on 1 January Bligh received two letters. One was from Macarthur, complaining again about Atkins and pointing out that if he sought to recover the debt by legal action, he would have to approach Atkins and ask him *to issue a writ to bring himself before himself.* Bligh referred Macarthur's letters to Atkins, who refused to comment as Macarthur was under committal for a trial before him.

Bligh also received a letter praising his administration signed by 833 of the inhabitants, mainly small settlers at the Hawkesbury. Bligh's opponents later claimed the petition had been organised by his supporters, such as Andrew Thomson, in response to the growing unhappiness of the colony's gentlemen. The small settlers – mainly ex-convicts – had every reason to support Bligh if they thought a showdown was looming between Macarthur and him. They had already shown their dislike of Macarthur when they rejected him as their spokesman at the time of Bligh's arrival in Sydney.

On 4 January Bligh's son-in-law, John Putland of the *Porpoise*, died of tuberculosis after a long illness. He was buried near the church with full military honours. The governor's mourning was interrupted by a third letter from Macarthur, threatening to appeal to London if Atkins was not put aside. Again Bligh refused.

Around 11 January, Macarthur moved into the Sydney residence of his friend Abbott, the nominal reason being that his eldest daughter, Elizabeth, was very ill and the doctor had suggested a change of air. According to Provost Marshal Gore, *From this period we are to date the active steps that were taking to wrest the administration out of the hands of Governor Bligh. What had hitherto been mentioned only in confidential whispers and conversations was now ventured to be more openly spoken of.* Preparations for the trial continued, with George Crossley assisting Atkins in interviewing the witnesses who would be produced against Macarthur, and drawing up a list of questions to be put to them in court.

Bligh's use of Crossley in preparing cases was roundly criticised by Macarthur and others, yet he had little choice. He had found Atkins in situ on his arrival and, appalled, had written to the Secretary of State describing the judge advocate in these terms: *He has been accustomed to inebriety; he has been the ridicule of the community; sentence of death has been pronounced in moments of intoxication; his determination is weak; his opinion floating and infirm; his knowledge of the law is insignificant, and subject to private inclination.* But Bligh had to work with what he had. Presumably he believed Crossley was better than nothing.

Now yet another dispute erupted. A lease on the church hill granted to Macarthur by Governor King had originally been intended by Phillip to be part of the churchyard. It also contained a well, previously accessible to everyone. Bligh moved to reclaim it, telling Macarthur he could select another piece of land as compensation. Macarthur nominated three other possibilities, but each was impractical. (It's not entirely unreasonable to suspect that he chose them for this reason.) Bligh suggested a fourth possibility, which Macarthur felt was unacceptable as it was *surrounded by all the vile and infamous characters of the town of Sydney*. On 12 January he told Bligh he would not budge from his lease, and had fence postholes dug around his block. The work was done by soldiers whom Macarthur employed as labourers when they were off-duty, but after it was well under way the men were ordered by Johnston – at Bligh's request – to stop.

The next morning Macarthur was standing on his plot of ground talking to Captains Abbott and Kemp and Lieutenant Bayly when Nicholas Divine, the superintendent of convicts rode up. He told Macarthur he had been ordered to take down any fence rail that was attached to the posts already in position. Macarthur, *thinking it necessary this should be decided*, put up one of the rails with his own hands. Divine got down from his horse and pulled off the rail. As he did so he said, in Macarthur's account, *When the axe is laid to the root, the tree must fall*. It was a somewhat perplexing line, presumably intended to invoke Luke 3:9: *And now also the axe is laid unto the root of the trees: every tree therefore which bringeth not forth good fruit is hewn down, and cast into the fire.* Divine went away and returned with a gang of convicts who pulled out all the posts and filled in the holes. They then removed all the fencing materials from Macarthur's lease, piling them some distance away. Given everything Macarthur had said about property in recent months, these actions were highly provocative, indeed political.

At this time occurred the first discussion that we're aware of specifically about deposing Bligh. In a letter to former Governor King, written after the rebellion, Captain Abbott said that Major Johnston discussed the possibility with him in mid-January 1808. It's possible there were previous discussions, but we don't know.

Macarthur now decided to obtain some popular support. He approached Sergeant-Major Thomas Whittle, a leading free spirit in the corps, who was to play an important role in subsequent events. Whittle had been involved in the mutiny on Norfolk Island in 1794 but came out of it all right, and later received a grant of 90 acres at Bankstown. By 1803 he had been promoted to sergeant and was known to be a Freemason. In the same year, when his son was beaten by a naval officer he had abused, Whittle caused a rumpus by threatening, on the parade ground, to cut off the officer's ears. We have already seen his claim that Bligh threatened to pull down his house, although Bligh later denied this. Now Macarthur said to him: *Sergeant Major, I am going to make a purchase of wine cheap; and if you will get a list of the non-commissioned officers that have families, and likewise of the private soldiers, that want a few gallons for their own use, I will purchase it for them at 5s. per gallon.* It seems to have been a good price for Cape wine, and Whittle made out the list and gave it to the generous Macarthur.

Macarthur continued to write to Bligh arguing that it was unfair for Atkins to sit in judgement on him. Morally Macarthur would seem to have had a case, but Bligh continued to reject it. Possibly he knew that Macarthur had purchased the bill purely to put pressure on Atkins, and was determined not to be manipulated in this fashion. Whatever his reason, he appears to have been blind to the way Macarthur was, however unfairly, manoeuvring him into a position where his actions could be made to look increasingly arbitrary and unjust to a growing number of people.

On 24 January 1808 the officers of the New South Wales Corps dined in their mess. Such dinners had once been common but had stopped in Governor King's time, and it had been decided, perhaps not coincidentally, to reinstate them now. Macarthur was not at the dinner, although some civilians, including his son Edward and nephew Hannibal, were. So too were Nicholas Bayly and Garnham Blaxcell, who had gone bail for Macarthur. He himself, according to one report, walked up and down outside enjoying the music. It was the birthday of the great Whig leader Charles Fox. Although Fox had died two years earlier, the anniversary continued to be celebrated with public dinners in London and elsewhere. We don't know if this influenced the timing of the corps' dinner that night, but many of those present would have been keenly aware of the day's significance. Macarthur's son James later recorded that his father was a *great admirer* of Fox.

According to Johnston, politics were not discussed at the dinner. One soldier, William Bremlow, remembered that *Captain Kemp was then in a state of intoxication: I saw Colonel Johnston take hold of the bottle, and holding it up in his hand, he said, 'Look here, Kemp; recollect tomorrow.' Then a parcel of the officers got up, and began to dance; I could not hear any more.* On his way home from this convivial event, to Annandale where he lived a quiet life with an ex-convict woman named Esther Julian and their children, Johnston overturned his gig and received various injuries. These confined him to his bed, when he finally reached it.

Bligh was later to assert that Johnston had this accident because he was drunk. But then, Bligh blamed a lot on alcohol. Johnston did not have a reputation as a heavy drinker, the regimental dinner had concluded at 10.00pm, and the colony's roads were poorly made and maintained and badly lit. There is no evidence this was Sydney's first drink-driving accident. However, by restricting his movements the next day, it was to contribute significantly to the events that now proceeded to unfold.

VIII

Rebellion

On 25 January the criminal court assembled in the Orphans House on Bridge Street for Macarthur's trial. This was climacteric: he had succeeded in pushing Bligh and the corps – all of them – out onto dangerous ground, where the outcome was uncertain and the possibilities dangerous. He had shown how he could use such opportunities for political advantage, and each time he had pushed the stakes higher with his rhetoric and claims against Bligh and Atkins. Events in New South Wales were now beyond the control of anyone except, possibly, the governor. But not this governor, whom Macarthur had baited with words and actions like dogs sent against a cornered boar.

The court consisted of Judge Advocate Richard Atkins, who at sixty-three was several decades older than most of the six officers on the bench with him: Captain Anthony Kemp and Lieutenants John Brabyn, William Lawson, Thomas Laycock, William Minchin and William Moore. Ordinary soldiers had been told to attend to support their leaders, suggesting strongly that Macarthur and the officers had discussed the matter and agreed on some course of action. Many civilians were also present in the courtroom, including Bligh's allies Robert Campbell, John Palmer and the Reverend Henry Fulton, who was the colony's chief chaplin while Marsden was in Britain on leave. Atkins swore in the officers and then, just as he was to be sworn in by turn, Macarthur asked for permission to protest against his involvement. Atkins refused this, claiming that until he was sworn in there was no court to hear the protest. The officers disagreed and allowed Macarthur to speak.

In retrospect, this denial of the judge advocate's authority can be seen as the beginning of the rebellion. Kemp later said they were aware that *a rancorous inveteracy subsisted between Mr Atkins and Mr Macarthur, which had displayed itself in a variety of shapes, we considered that Mr Atkins was ineligible to sit as a juryman on Mr Macarthur's trial; and when we reflected that Mr Atkins was the magistrate who had committed Mr Macarthur for trial, and that Mr Atkins ... was to sit as prosecutor, judge and juryman, combining the three characters in one, we were confirmed in our opinion, that upon every principle of justice and rectitude he ought not to sit.*

Macarthur proceeded to launch an extraordinarily vituperative – if not unprecedented – attack on Atkins. He began by claiming to be unaware of the charges against him. This was a standard Macarthur tactic in such situations, but this time it was true, as Atkins had several times refused to tell him what he was to be charged with. Macarthur then recalled how he'd applied to Bligh for an alternative judge advocate to be appointed to hear the case, and how his request had been rejected. *Defeated in this attempt to obtain what I know to be right,* he went on, *my only alternative is to resort to the members of this court; and I do so under an entire confidence that what I can prove to be my right, you, as men of honour, will grant me.*

Macarthur informed the men of honour that Atkins was unfit to sit in judgement on him because he owed him money. Furthermore, the two had been involved in disputes in the past, as a result of which Macarthur was the subject of Atkins' *vindictive malice*, expressed through *the propagation of malignant falsehoods, and every act of injustice that can be expected to proceed from a person armed with power, against a man whose life and conduct is, I trust, a public satire on his own.* Macarthur was also upset because the plan of prosecution against him had been prepared by *that well-known dismembered limb of the law, George Crossley*. His evidence for this was a memo in Crossley's handwriting which, Macarthur claimed, had fallen from its writer's pocket and been found and brought to him. *This precious document*, he claimed, *came into my hands as if by the interposition of Divine Providence.* (Actually it was obtained for him by an Irish man who'd been drinking with Crossley.) As his address to the six officers approached its end he gave a string of legal references, suggesting he owned a good library, followed by the words: *Now, gentlemen, for God's sake remember you have the eyes of an expecting public upon you, trembling for the safety of their lives, liberties, and properties.* Here as elsewhere in his public pronouncements, Macarthur showed the influence of the thinking of Locke and the style of Edmund Burke, the great liberal who had died in 1797. (See Appendix for a copy of the speech.)

The courtroom was full, although no building in the colony could have contained the thousand people Atkins later claimed were there. This was the best theatre available in Sydney, and the audience was not to be disappointed. Macarthur concluded with some stirring sentiments: *To you has fallen the lot of deciding a point which involves perhaps the happiness or misery of millions yet unborn; and I conjure you in the name of Almighty God, in whose presence you stand, to consider the inestimable value of the precious deposit with which you are entrusted.*

Macarthur was not previously known for his interest in the unborn millions. As for Almighty God, the Reverend Fulton, who knew and disliked Macarthur, noted that he believed *the only use of religion was for political purposes*. However, the speech impressed many of the others who heard it.

I will commit you to gaol, sir, cried out Atkins.

You commit, sir! No, sir, I will commit you, Kemp responded.

The soldiers cheered and there was a fierce argument for three or four minutes, after which Atkins quit the room, claiming that without him there could be no court. He yelled out that the crowd should disperse, but Kemp and the other officers cried out, *Stay, stay! Tell the people not to go out, we are a court*. This was a considerable presumption on Kemp's part. Macarthur then addressed the officers again, saying *Am I to be cast forth to the mercy of a set of armed ruffians – the police?* He claimed to have received private information that he would be taken the moment he left the building, and requested a military escort. This he received, and was released on bail.

Atkins had left behind his papers and later asked for them; the officers refused. Like Macarthur before them, they applied to Bligh for the appointment of another judge advocate so the trial could proceed. The governor denied the request, and sent for Major Johnston, who was in bed at home. Johnston returned a verbal message that he was too sick to travel the four miles from Annandale to Sydney, indeed too sick even to write a reply. He had an injured right arm.

Gore and the four magistrates at Government House believed the officers were not a court, and therefore had no right to release Macarthur on bail, and the magistrates ordered that he be arrested, to which Bligh agreed. The next morning this occurred and the richest man in the colony was taken and thrown into the common gaol. By the standards of the day, in Britain and New South Wales, this was an insulting and provocative action on Bligh's part. Gentlemen rarely treated other gentlemen in this manner, whether they were legally entitled to or not. John Macarthur employed about forty free men. There were another thirty convict labourers, as well as thirty wives and children of the paid employees; altogether a large establishment producing a substantial income. He had already sold some 6,000 sheep and 300 cattle and was now capable, he said a few years later, of selling 100,000 pounds of meat annually. He was the representative of the colony's establishment.

One can imagine Macarthur's emotions on the morning of 26 January 1808 on being seized by the constables and greeted at the gaol entrance by a warder who was a former servant whom he had discharged from his house for stealing. Shortly afterwards he encountered another guard, also a former employee, whom he had sacked for being associated with a gang of cattle thieves. Fortunately the deputy-gaoler was a decent man Macarthur had long known, who promised to look after him. He believed that a group of men planned to kill Macarthur, but (as Macarthur's son James later relayed the story) *there was a cutlass concealed in the cell for his honour and he had one for himself, and they would sell their lives dearly*. In the event, such sacrifice was not called for.

At Government House Bligh had assembled a group of magistrates and officials. These were Richard Atkins, William Gore, Robert Campbell, John Palmer, the Reverend Henry Fulton, James Williamson and probably the surgeon Thomas Arndell. This group included pretty well all the gentlemen who supported Bligh in the colony. Most were just as involved in trade as the officers: Palmer and Williamson had abused their control of the government store for many years to enrich themselves. Sometimes they had acted in alliance with the military officers in this and other commercial ventures. Robert Campbell had been about to enter into a business scheme with Lieutenant Putland when the governor's son-in-law died. Fulton had his own reasons for disapproving of Macarthur, whom he believed to be a Deist who *did not believe in revelation in express words* (i.e. in the Bible or organised religion). George Crossley was also there. Bligh now received another letter from the six officers, asking that Macarthur be released. Crossley suggested that Atkins present a memorandum to the governor indicating that the six officers were guilty of *a usurpation of His Majesty's Government* and of inciting *treason*, and this was drawn up and presented to Bligh. The governor then issued summonses to the officers to appear at Government House the next morning to answer the charges *of certain crimes*.

When the officers received these, they were greatly disturbed. Kemp *concluded that under the influence of the notorious George Crossley a summary tribunal would be resorted to, composed of the dependent magistrates, and that the members of the court would have been ignominiously ordered to gaol*. A soldier later recorded the conversation between some of the officers after they had heard that Bligh was accusing them of treason.

Laycock: *By God we are all going to be hanged.*

Unnamed: *Why so?*

Laycock: *Old Bligh has summoned us to appear before him.*

Minchin: *If you are afraid I am not, and will find a way to cool him.*

Laycock: *How is that?*

Minchin: *By arresting him before he arrest us.*

Minchin had been particularly disturbed by the way Bligh

had used his bench of magistrates the year before to punish Michael Dwyer, the Irish rebel, after the court had found him not guilty. Minchin's farm at Cabramatta was close to Dwyer's and he thought him an industrious and deserving settler. He'd pitied the way Dwyer's family had had to exist on charity after Bligh had him transported. So thoughts of the governor's ruthlessness and abuse of the court system would have been much in his mind now; his fear that Bligh might arrest them was genuine. It's also worth recalling for a moment the clash between the court and Governor King after the trial of Lieutenant James Marshall in 1801. In that case the court had refused the governor's request for them to reconsider the matter, and the naval lieutenant James Grant had actually told his fellow judges he would not be surprised if King put them in gaol. This is further evidence that the fears of the officers in 1808 were genuine.

Now Laycock protested that to arrest Bligh would be mutiny. Minchin said he intended to summon Johnston, and the others agreed. Lawson said: *That is the best thing to be done.*

Bligh himself sent another message to Johnston, asking that he come to Sydney or send his deputy, Captain Abbott, from Parramatta in his place. The language of this letter is essential in understanding Johnston's subsequent actions. Bligh informed him that the judge advocate had *laid a memorial before me against six of your officers for practices which he conceives treasonable*, so he had summoned the officers and magistrates to appear the next day. Johnston later claimed that he understood the letter to mean Bligh, using Crossley as an adviser and Atkins as his tool, would in effect try the six officers for treason before the magistrates, who were all his allies. The case of Dwyer indicated that Bligh was prepared to override the courts. Johnston did not believe the officers would be hanged, but he did believe they might end up in gaol. (Macarthur's example showed Bligh was happy for the colony's gentlemen to be imprisoned alongside common felons.) In any event, the mere fact that Bligh had dared to allow the officers to be accused of treason would have suggested to Johnston that the governor had now gone too far. Should he try to reason with Bligh? In principle, yes. But from his experience of dealing with the governor during the past two years, he had formed the strong view that Bligh could not be reasoned with, that he could not be trusted to act like a gentleman. Such were the elements of Johnston's dilemma.

While the prospect of having his officers gaoled would have horrified Johnston, so too would the likely implications of such an action. He later said he believed the ordinary soldiers would have rioted, and possibly even have killed Bligh, had most of their officers been locked up. This was derided by Bligh and his supporters as implausible. But it was a realistic fear. We have already seen how the men of the corps, while possibly no worse than those in the rest of the British army, could be rebellious and violent. Riots and mob action had been frequent in Britain for decades – there had been riots over football, food shortages, turnpikes, enclosures and political issues. There had been mob attacks on the persons and properties of politicians and magistrates, accompanied in some cases by the burning of effigies and the illumination of windows at night to demonstrate support for the riots. (Unlit windows were smashed.) So, while it might have been arguable whether such things would occur in this case, for Johnston to believe them possible (and after all, he knew his men better than Bligh did) was quite rational.

By this stage the town was seething with rumours and tension. It was not until 5.00pm that Bligh received another verbal reply from a messenger, to the effect that Johnston remained too ill to write but would have someone pen a response on his behalf later that evening. The message, following those earlier in the day, indicated that communications between the governor and the head of the military had broken down. The dispute was now out of anyone's control. Macarthur had successfully pushed Bligh into a situation where his actions were widely seen as terrifyingly unpredictable.

The governor sat down for dinner with his daughter, Griffin, Campbell, Palmer, and Williamson; presumably none of them was particularly happy about the way things were going. Bligh and Mary were dressed in black, in deep mourning for the recently departed John Putland. What the diners did not know was that Johnston had been alerted earlier to what was going on by his doctor, John Harris, who informed him that the populace was shocked by Macarthur's gaoling and that *an insurrection of the inhabitants was to be feared.* Johnston had been brooding on Bligh's apparent intention of arresting six of his officers for treason, which would have left only two to run the regiment in Sydney, and he was genuinely appalled that Bligh suffered *himself to be guided by a wretch like that man Crossley to prosecute and oppress His Majesty's subjects.* At last, despite his injuries, the major, his face bruised and his right arm in a sling, decided to act. He hastened into Sydney by carriage. Adjutant Minchin, whose conversation with his fellow officers we overheard, was at the reins as they approached the town. Their course would have taken them along what is now George Street and past Bridge Street. It was here that Johnston should have ordered Minchin to turn right and take him to Government House, to discuss the matter with Bligh. They were the two senior men in the colony, and they should have been able to sort this out. That Johnston felt he could not do this tells us a great deal about the relationship that now existed between the two men.

Some later writers, such as H.V. Evatt, suggested Johnston

was manipulated by the Machiavellian Macarthur. This fits in with a view of the story as a conflict between pure good and pure evil, in the form of Bligh and Macarthur. However, this version makes sense only if Johnston is seen as a mere cypher, yet there is no justification for this view. He was an experienced officer who had fought in battle, led his men successfully for many years in a distant posting, capably put down the 1804 Vinegar Hill uprising, and publicly criticised Macarthur only a few months earlier, during the magistrate's hearing. That he might also have discussed with Macarthur the prospect of getting rid of Bligh subtracts nothing from the picture of him as a competent and cautious officer.

Johnston later told his court martial that, even if he had talked to Bligh and they had come to an arrangement, the governor's *sacred and solemn pledge as an officer and a gentleman might have been insufficient security*. But in any case he believed it impossible to negotiate with William Bligh. When he had tried to discuss the important matter of the roster for officers serving on the court, Bligh had *displayed the utmost anger; dismissed me roughly from his presence; and within two hours removed Mr Jamison, who had merely accompanied me to witness the interview, from the situation which he held as a magistrate*. This belief – that Bligh was not a man of honour, someone you could negotiate with in good faith – was strongly held by the rebels and vital to the course of action they adopted. It was not just that Bligh did bad things: he was a bad person.

Johnston lived a pleasant life, as far as we know in a state of domestic harmony with Esther and his children. His material circumstances were superior to those he would be likely to experience if ejected from New South Wales. He would have been reluctant to put all this at risk, so it is extremely unlikely that at this juncture his actions were determined by Macarthur, who was in gaol. It is far more likely that he was greatly disturbed by Bligh's assault on the honour and stability of the regiment for which he was responsible. He saw this as the latest in an escalating series of such assaults, and at this point must have decided Bligh had crossed the line of reasonableness and honour. Only such a belief would have forced Johnston to put his future at risk and act as he now did: by not ordering Minchin to turn right, Johnston committed himself to a mutiny.

As he entered the centre of Sydney, Johnston found *the common people ... in various groups in every street, murmuring and loudly complaining, whilst others were watching the movements of Crossley and the magistrates who frequently passed from the Judge Advocate's to Government House.* Most of the ordinary soldiers of the corps were married, or at least cohabitating with women, and lived outside the barracks and mixed on a daily basis with the general population. About two hundred of them were former convicts. As well as this, a large number of soldiers were discharged in the colony – by 1810 the number was 422, or 17 per cent of the male population. Then there were their teenage children, some of whom by now could have been eighteen years old. Therefore, to have the soldiers involved in a controversy meant that a critical mass of the other inhabitants of Sydney was also involved. And, as many as two-thirds of the soldiers – a high proportion by British standards – owned property, whether land, houses or cattle, and were therefore concerned by Bligh's attacks on the security of houses and leases.

On Johnston's arrival at the barracks by the old parade ground he consulted with his officers and some of the leading free settlers, including the Blaxland brothers. *They solemnly assured me, if I did not [arrest Bligh], an insurrection and massacre would certainly take place.* It was put to Johnston that, the way things were going, Bligh would soon be assassinated, so that by arresting the governor he would actually be saving his life. Johnston was told that Macarthur had been gaoled, *and there was much reason to fear he would be privately made away with.* He ordered that Macarthur be released from prison, signing the letter *Lieutenant-Governor*, a position that could not exist while Bligh was in charge. This was the point at which Johnston effectively assumed the government of the colony for himself.

How did these men view rebellion against authority in a general sense? It is important to be aware that Georgian Britain condoned mutiny by gentlemen, at least more often than we might imagine. While three men were hanged for the mutiny on the *Bounty*, they were not gentlemen. Britain in fact had a tradition of revolt, and most Englishmen were proud of Magna Carta and the civil wars of the seventeenth century, which they regarded as part of the mainstream of their nation's progress. The British monarchy was descended from William and Mary, fairly popular replacements for James II in the Glorious Revolution of 1688. Ever since, the people had been suspicious of royal power. George III – monarch for the events covered by this book – spent his reign fighting parliament to maintain his dwindling power, and was widely resented for doing so. The decision by the American colonies to rebel (like the British civil wars, a rebellion primarily of the gentry, not the people) was supported openly by many gentlemen in England, including several dozen aristocrats. It was widely considered that George had driven the Americans to rebel through his excessive exercise of authority. Charles Fox wore a blue frockcoat with a buff waistcoat (the American colours) for the rest of his life. So the English were suspicious of anyone who had too much authority, and therefore had a certain tolerance of mutiny and rebellion by gentlemen, if they believed it justified.

In 1776 the British members of the council of Madras, all gentlemen, deposed and imprisoned their governor, Lord Pigot, who died in captivity. The only punishment meted out was a fine of £1,000 each for the four ringleaders. Warren Hastings, a former governor of Bengal, was hounded in 1786 for various alleged offences by parliamentarians including Burke and Fox. Hastings was impeached and, even though he was eventually cleared, the very public trial implanted notions of the accountability of leaders firmly in the public consciousness. In 1806 Governor Thomas Picton of Trinidad was tried and found guilty of using torture to extract a confession from a prisoner.

The details of these cases differed from what occurred in New South Wales in 1808, but I mention them because they suggest the importance, as our story approaches the Bligh rebellion, of not judging it according to today's standards. The historian James Johnston Auchmuty has shown that gross disobedience and even mutiny in the British armed forces were not unusual, even at high levels, in the eighteenth century: *The spirit of disobedience, for cause shown, the feeling for one's rights, ran through the whole of society from youth to age*. One aspect of this that he points to was the phenomenon of school mutinies. After the Harrow School Mutiny of 1771, the future Marquess Wellesley was expelled – and promptly accepted by Eton. During the Winchester School Mutiny of 1793 the Buckinghamshire Militia had to be called out. Anyone whose view of Macarthur's world is impeded by the mammoth edifice of Victorian England, and the efficient and effective power structures it established, needs to try to see around it and back to the wilder world of the Georges.

In Sydney on the evening of 26 January 1808, Macarthur was released from gaol and arrived at the barracks where he found about seventeen soldiers and civilians in a room with Johnston. Macarthur went alone with Johnston into an adjoining room. He suggested that a letter be prepared and signed by the civilians present, calling on Johnston to arrest Bligh and take over. The letter was promptly drawn up by Macarthur himself, and read as follows.

Sir, – The present alarming state of this colony, in which every man's property, liberty and life is endangered, induces us most earnestly to implore you instantly to place Governor Bligh under arrest, and to assume the command of the colony. We pledge ourselves, at a moment of less agitation, to come forward to support the measure with our fortunes and our lives.

It was signed first by Macarthur and then by over 150 others, most of them after Bligh's arrest. Those who signed beforehand were, in order: John Blaxland, James Mileham (a surgeon), Simeon Lord, Gregory Blaxland, James Badgery (free settler), Nicholas Bayly, Garnham Blaxcell and Thomas Jamison. None of the military officers signed.

Bligh later claimed that Macarthur and Nicholas Bayly had persuaded Johnston to rebel, but there's no evidence of this. Macarthur and Johnston were not close, and as we have seen, Johnston had claimed publicly that Macarthur was a trouble-maker. We do know that Johnston and Abbott had discussed it beforehand. The conversation between the officers quoted earlier suggests Minchin had reason to think Johnston would agree to arrest Bligh.

However, all we know for certain about Macarthur's involvement is that the writing of the above letter is the first moment at which he clearly aligned himself with Johnston. At Johnston's trial he gave this explanation for his action: *I conceive the arrest [of Bligh] to have been occasioned by the dread that was entertained of the six officers being sent to gaol, and of the resentment that would have excited among the soldiers and the inhabitants, the dread that was entertained of an insurrection; such were my own motives.* A key element in the decision was the small number of officers in the corps. Major Abbott was at Parramatta, Johnston was unwell and Lieutenant Draffin was on duty but mentally unstable. Had the six officers been imprisoned, this would have left just Johnston and one other capable officer in Sydney, Archibald Bell, to keep control of three hundred soldiers.

And so the decision was made. Bligh had given the rebels a range of reasons to dislike him, and had affronted the sense of honour of Macarthur and the military officers, and some of the other inhabitants, by his actions and by his words. Often this had been done by their style and tone as much as by their content. It was what he had done on the *Bounty*: he had insulted other gentlemen from behind the defence of his superior position, so they could not call him out. He had pushed them too far, by the standards of their class and their era. Twenty-five years earlier, Dr Johnson had said, in defence of the duel, *a man may shoot the man who invades his character, as he may shoot the man who attempts to break into his house.* Bligh had invaded the characters of the six officers, and therefore of the entire New South Wales Corps, by accusing them of treason.

Once the dispute became personal, each side raised the stakes, and assumed the worst of its opponents. Lieutenant William Minchin later recalled that many people initially *had expressed their fears for their own persons and properties, from Mr Macarthur, who was a man of property, being committed to prison.* Then Bligh had upped the ante, and Minchin recalled *I believe the general alarm arose from the idea of the officers being committed to prison.* The officers, after all – and unlike William Bligh – represented the continuity of authority for most of the colony's existence. They were its establishment. So, Minchin said, *If the six members [of the court] had appeared before the Gov-*

ernor, I have not the smallest doubt but that they would have been committed to prison until His Majesty's pleasure was known; and had the officers been committed to prison, there certainly would have been an insurrection; the gaol would have been pulled down, [by the soldiers] and where it would have ended I cannot pretend to say.

As the sun set, most of the regiment, with bayonets fixed, marched from the barracks into Bridge Street and over the Tank Stream towards Government House. The regimental flag was carried by the ensign and the fifes and drums were playing 'The British Grenadier':

Some talk of Alexander, and some of Hercules,
Of Hector and Lysander and such great men as these,
But of all the world's great heroes, there's none as can compare,
With the fol-de-rol-de-rol-de-rol of the British Grenadier.

Sergeant-Major Whittle urged on his men, and yelled solicitously to some kids at their heels: *Children, go out of the way, for some of you I expect will be killed.* George Suttor, a farmer and Bligh supporter, later recalled hearing Macarthur call *Hush! Hush!* to the troops, suggesting that he was aware of the need for the rebellion to occur in a restrained, indeed an honourable manner, that would look good when it came to be reviewed in London.

Inside Government House, dinner on the ground floor had finished and the governor and his guests were relaxing over port when the news of Macarthur's release was delivered. Bligh immediately left the table and went upstairs to put on his uniform. Soon he would have been informed of the approach of the regiment, and have heard their music. He called for his horse, hoping to escape and reach the Hawkesbury, where he knew the settlers were friendly. He began to collect some papers, and then went into a small back room upstairs and proceeded to tear up certain documents, giving the pieces to his servant, John Dunn, to destroy. Other papers he hid beneath his waistcoat.

Gore arrived with a copy of Major Johnston's order for the release of Macarthur. Mrs Putland's screams could now be heard, from the front gate where she was trying to stop the soldiers. John Palmer, who had gone out to assist her, ran back inside and closed the front door, which had a large piece of glass in its top half, as the soldiers made their way up the path and surrounded the building. The Reverend Fulton refused to open the door to one lot of soldiers, led by Captain Kemp and Lieutenant Lawson, telling them they might drive their bayonets into him if they chose. According to Fulton, *Lieutenant Lawson then said, that Captain Kemp would be obliged to me if I would tell the governor he wanted to speak to him.* But another group of soldiers came in the back way, and Palmer persuaded Fulton to open the door, as he was *fearful they would do him some mischief by driving a bayonet through the glass.*

Surprisingly, it took the soldiers up to two hours to find Bligh. During this period they searched the house several times, followed by the outhouses and the grounds (it would have been dark by then), even sending soldiers out along the roads in case he had made his escape. Sergeant Major Thomas Whittle was an enthusiastic searcher, yelling out that Bligh *was gone down to Palmer's.* Troops were sent to the commissary's house and searched it without success, and returned to Government House.

Eventually an infuriated Whittle yelled out *Damn my eyes, I will find him – soldiers, come up stairs again, I will have another search.* Bligh was then discovered when one of the troops felt beneath a bed in a servant's room, and the governor was hauled out with his uniform covered in dust. According to Private John Sutherland, the Governor was *under the bed, sup-*

porting himself forward upon his two hands, with one foot placed against each of the two posts, either on one side or at the bottom of the bed with his back pressed upwards against the bottom of the bed. I saw the cover of the bed, which was hanging down, move ... I put my piece [gun] under the bed, and moved it along, to see if there was any thing. I found nothing under the bed [at first] ... The second time I put my piece under the bed, the piece struck the boot; the boot went from the post of the bed, and made a slip on the boards. According to Sutherland's account, Bligh must have wriggled out on the far side of the bed when discovered, because *I put my hand across the bed to help him over it.* This raises the question of whether a bed could be low enough to perform this manoeuvre but high enough for Bligh to hide under it with an arched back in the ingenious manner described.

Bligh later claimed he had been standing in the room all along, tearing up documents. But it seems impossible that he could have done this and not been found for such a period in what was not a particularly large building. In any case, once he was discovered a soldier, seeing the governor's hand fiddling with the papers under his waistcoat, presented his bayonet and cried, *Damn your eyes, if you don't take your hand out of there, I will whip this into you immediately.* To which Bligh responded: *Sergeant, keep the man off, I have no arms.* Lieutenant William Minchin appeared at the door and, in his version, was told by one of the soldiers, *Take care, he has got a pistol.* Minchin thought Bligh was just tucking his shirt frill into his waistcoat, which was open. He ordered the soldiers out, and one of them told him as he was leaving, *We found him there, sir,* pointing under the bed. Minchin saw that Bligh's coat was covered in dust and feathers, and that the Governor *appeared to be very much agitated; indeed I never saw a man so much frightened in my life ... When I went into the room he reached his hand to me, and asked me if I would protect his life ... he took my arm, and we walked down stairs together.* They passed Surgeon John Harris, who later recalled that Bligh *seemed extremely warm, in a violent perspiration.* He also noted that there were *two bunches of feathers on his coat.*

Bligh was to say that at this moment he expected to be murdered, although he might have been laying it on to arouse sympathy. He demonstrated courage in other situations during his career, including the *Bounty* mutiny, and it could be that the rebels manufactured this incident to embarrass him. But we know from Bligh's own account of the evening that the rebels reached Government House about sunset, or maybe a bit later, and left at 9.00pm. This means they were there for several hours, and there is nothing in his account or anyone else's to explain what they did that would have taken more than half an hour or so, had they found him soon after they arrived, which is what he implied.

In favour of the rebels' story is that it emerged immediately from several witnesses, both officers and men, who were consistent over several years of repeating it. My guess is that Bligh did hide under the bed, but was motivated not by cowardice but by a desire to escape and make his way to the Hawkesbury. When he saw how this was being interpreted and accepted as cowardice, he decided to lie about it.

On the ground floor Bligh met his daughter and had a few words with her. Then he was handed a letter from Johnston telling him that *you are charged by the respectable inhabitants of crimes that render you unfit to exercise the supreme authority another moment in this colony.* Johnston wrote that in taking over he was *acting on the advice of all my officers, and by the advice of every respectable inhabitant in the town of Sydney.* Then Lieutenant Moore advised Bligh that Johnston wanted to speak to him, and the Governor was led into an adjoining room. This was full of people, in Bligh's words *just like a Robespierrean party, or a revolutionary tribunal.* Apart from the soldiers there was a group of free settlers, including John and Gregory Blaxland, and Edward and Hannibal Macarthur. John Macarthur kept himself outside with the bulk of the troops, presumably already aware of the need to minimise his role as far as possible. Others who were with him included D'Arcy Wentworth, Simeon Lord and Henry Kable.

Now Johnston confronted Bligh and repeated the gist of his letter. The governor's papers and the great seal of the colony were confiscated, and he was placed under house arrest.

Major Johnston, Bligh said, *I suppose you will not take my secretary away, he may remain with me?*

Oh yes, certainly, Johnston replied, and then had a second thought. *Mr Bayly, go and ask the inhabitants outside whether they approve of Mr Griffin's remaining here.*

Bayly went out and returned with the information that the mob did not approve of this arrangement, so Edmund Griffin was taken away. (According to Bligh's version, Bayly objected and went out and asked Macarthur, who supported him.) At 9.00pm Bligh was left in the house with just his daughter and Mrs Palmer, with five soldiers to guard them. Johnston had declared martial law and placed a roadblock on the Parramatta road.

Outside in the town there was much rejoicing; bonfires burned and the citizens celebrated and drank to the overthrow of the tyrant. But the twentieth anniversary of the first settlement of Australia was observed more soberly by the leading rebels, who proceeded to work their way through the governor's papers and interrogate his leading supporters. On that night Edmund Griffin was hauled before what Fulton described as *a kind of revolutionary committee,* in Johnston's town house. He was taken by Lieutenant Brabyn and John Blaxland,

and the committee consisted of Blaxland, Townson, and Lieutenant Draffin. Johnston stood by. Macarthur arrived and joined in the proceedings, asking most of the questions, which were designed to extract information harmful to Bligh regarding the governor's use of the public stores.

Macarthur had helped bring about the most dramatic event in the early history of Australia. Honour had been challenged; honour had been satisfied, in a piece of tense but bloodless theatre involving weapons, similar to the *Bounty* mutiny and to most duels. But in this case, unlike in a duel, the matter was not concluded and the danger was not over. For one party, it was just beginning.

Why did Macarthur bring on the Bligh rebellion? The preceding account suggests that the question might be misleading: it assumes he had both a plan and the means of implementing it, whereas it is far more likely that he simply had a great deal of rancour but possessed limited control over the precise direction taken by events as they unfolded. Events often look inevitable in retrospect, yet, but for Bligh's decision to call the six officers before him, the rebellion might not have occurred.

There is no doubt that Macarthur and others wanted to hurt Bligh and did a great deal to damage him in the period before the rebellion, both in New South Wales and Britain. When news of Bligh's deposition reached London, on 20 September 1808 the Honourable Charles Greville wrote to Sir Joseph Banks: *I fear the plans against Bligh have been extensively laid and artfully conducted. Certainly prejudice has been more barefaced than in any case I ever saw.*

Yet the rebellion itself occurred only because of a tortuous chain of events impossible to have predicted or engineered. My belief is that Macarthur provoked Bligh to act rashly, but without knowing what form this would take, in the hope it would lead to his recall. (As we've seen from the evidence of Edmund Griffin, Bligh was acting towards Macarthur similarly.) Macarthur hated Bligh because he threatened his interests and his status. But by gaoling Macarthur and accusing the officers of the New South Wales Corps of treason, Bligh turned this political dispute into an affair of honour; he made it far more personal than it had been before. This introduced a qualitative difference into the running battle between the governor and his powerful opponents, who now sought, as was expected in an affair of honour, a quick and, if necessary, violent resolution. For the good of the colony no one could challenge the governor to a duel, so what occurred next was a version of a duel, in that it restored the self-esteem of Bligh's opponents and relieved the pressure that had built up.

It was now up to the rebels to persuade the world that their action had been necessary, and conducted in an honourable manner.

IX

Aftermath

While the rebellion was occurring in Sydney, there were rumours and press reports in London that Bligh was to be recalled by the government. It seems likely that Macarthur and perhaps other colonists had written to England in the previous year urging their allies to agitate for the governor's return. Elizabeth Bligh told Sir Joseph Banks that there was *so strong a party, from interested motives, making every effort to recall and ruin Bligh in his character … I find the party have been cruel enough to get somebody to represent to the Admiralty … that Bligh, during the scarcity in the colony, sold provisions at high prices for his own emolument*. Part of the attack involved claims in the press that under Bligh the colony was in dire straits. In November 1807 the Whig newspaper the *Morning Chronicle* announced that Macarthur *has been forced, with a heavy heart, to abandon in a great degree his experimental farm, for the improvement of the breed of sheep.* The appearance of this article would have surprised Bligh, although not, one suspects, Macarthur.

In February 1808 Captain Joseph Short, who had crossed Bligh on the journey to Sydney and been sent back to England to answer charges including drunkenness and unofficer-like conduct, was tried. The court honourably acquitted him of all charges, despite the fact that Philip Gidley King, John Oxley and John Houston – the three officers who had found against him back in Sydney – were all available in England. None of them was called as a witness. Rear Admiral Isaac Coffin, the court's president, took the unusual step of informing the Admiralty that Bligh had been influential in bringing the matter to trial, and as a result, Short *has been deprived of the command of his ship; precluded from benefiting by the benign intentions of government in locating a tract of land ordered to be granted to him; obliged, with a wife, and six children under twelve years of age, at an inclement season of the year, to return to this country in a leaky ship, whose distresses occasioned the loss of his wife and one child.* He had also been financially ruined by the experience, and it was all the fault of William Bligh.

Elizabeth Bligh, along with many former inhabitants of Sydney, followed the court martial closely and reported that it was turned into an attack on Bligh, who in fact had had almost nothing to do with the charges or the events involved. *When Captain Short began his defence they listened with every complacency and sat for a long time to hear him defame and belie the governor of New South Wales, who was no party to the trial, and had nobody to defend him.* But Coffin wanted Bligh recalled so his friend Captain Kent might replace him. *Kent went down to Portsmouth [for the court martial] in the same chaise with King*, Mrs Bligh wrote, *and at Hunter's lodgings with Short planned the business – Foveaux was very active against you … Foveaux is very ill-disposed to you and I hope you will send him to Norfolk Island.* (In fact, Foveaux was soon to return to the colony as the intended lieutenant-governor of Norfolk Island.) It was quite a conspiracy, which involved King, Oxley and Houston reversing (or at least remaining silent on) the opinion of Short's actions they had reached in Sydney, and which also drew in John Hunter on Short's side. Elizabeth informed Sir Joseph Banks that another scheme was *afoot to give General Grose the governor's job.* She noted that the *Morning Herald* of 1 February actually announced Bligh's recall. There was no basis for this.

It is, incidentally, interesting to see that so many people wanted the demanding and thankless job of governor of New South Wales.

Sydney, 1808

For several days after 26 January the celebrations continued. According to evidence later given by John Palmer, armed soldiers ordered the populace to collect firewood and make more bonfires to celebrate Bligh's overthrow. Chief Constable Oakes later said that at Parramatta there was a general illumination and bonfires, and sheep were roasted. Oakes himself was burned in effigy (which word also meant a portrait), forcing him to move his family to a friend's house *as the people who were concerned in this seemed to be very much enraged against me.* In Sydney, according to the memory of a soldier named Isaac Champion, Sergeant-Major Whittle drew a picture, claiming that he *had got the old tyrant and old villain's picture drawn in a proper manner, and that he would expose it. … It was placed in his room, with a lamp on each side, for the public inspection of any person who thought proper to go there.* The picture was of Bligh being dragged from under a bed by two privates. It was small and done on canvas. On the evening of the 28th, Champion saw an effigy burned in a bonfire *and there were three loud shouts from the people in general; and at that time there was a number of officers and ladies walked by ... I particularly observed Johnston, Minchin, Major Abbott, a gentleman of the name of Macarthur, and some ladies.* The military band was there, and as the effigy burned they played a tune called 'The Silly Old Man'.

Sergeant-Major Whittle told people he would have placed his effigy on top of his house except that Johnston had told him *it would hurt him in England.* When Whittle was organising the building of a bonfire, Johnston had to tell him to move it, as it stood on the church hill and was too close to the tomb of Lieutenant Putland. On a public house near the wharf a sign was hung representing the figure of a Highland soldier, sword drawn, standing with a snake beneath his feet and the words *The ever-memorable 26th of January 1808!* It hung there for several weeks. In Sydney and the Hawkesbury, petitions were drawn up and long lists of signatures obtained, sometimes by force, in support of Johnston's action. Signs appeared around the town with phrases such as: *No Gore, No Tyranny,* and *Johnston for ever.*

While the populace celebrated long and hard, the leading rebels continued to work with a puritanical zeal to gather evidence that would show the British authorities they had acted with justification. They appear to have been acutely aware that their behaviour in the days following the rebellion would be closely noted and contribute to how their actions were assessed when the day of judgement came. As we have seen, duels had to be conducted with scrupulous care, according to the code of honour, in case they ever came to court. This rebellion was no different. So Macarthur avoided public celebration. He didn't want another Mackellar saying he had *exulted* in victory.

The interrogations of the magistrates and other allies of Bligh went on, probing for evidence of the governor's crimes. Essentially what they found was that Bligh had acted rudely to many people, he had abused his position in minor ways to make money from his farm on the Hawkesbury, and some of his officials – like those of earlier governors – had also abused their positions to make money, although he might not have been aware of this. He had been mean with land grants, but, for reasons largely outside his control, there were no convicts to open up new land anyway. He had probably been a far more effective governor than Hunter or King in defending the interests of the many small farmers against the traders and the big landowners, but of course his interrogators had no sympathy with this aspect of his regime.

Those interviewed by the rebels included Bligh's dinner guests from the night of the 25th and various deputy commissaries and storekeepers. On 27 January Campbell, Palmer and Atkins were dismissed from their positions, as were all the magistrates and William Gore. Even the Reverend Fulton was sacked soon after. The magistrates were replaced by the officers of the corps and their allies. Johnston issued a proclamation assuring the people that *In future no man shall have just cause to complain of violence, injustice, or oppression. No free man shall be taken, imprisoned, or deprived of his house, land, or liberty, but by the law. Justice shall be impartially administered.* This announcement was the high point reached by revolutionary rhetoric. It reminds us a little of the American Revolution, almost the only aspect of the Bligh rebellion that does.

On 28 January Nicholas Bayly, Johnston's newly appointed secretary, signed a letter to Bligh informing him that the former magistrates had confessed that their old boss had *been acting upon a settled plan to subvert the laws, to terrify and influence the courts of justice, and to deprive every person who had the misfortune to be obnoxious to you of their property, liberty and lives.* Johnston thought Bligh should know this so he could prepare his defence, and promised to provide him with transcripts of the interrogations. The letter, obviously composed with an eye to future readers, closed with Johnston's wish to do everything he could to alleviate the distress of Bligh's *present situation.*

In fact the magistrates had not said this at all, and when Robert Campbell saw this allegation he too put his thoughts down on paper in a letter to (and presumably requested by) Bligh. *I beg leave to express my indignation and contempt at so palpable a falsehood,* he wrote, *invented, I am convinced, for to deceive the British Government, and to afford the officers of the New South Wales Corps a renewal of their disgraceful and degrading practices your wise and judicious regulations had so much curtailed.* He believed the rebellion had occurred because Bligh had been successful in curtailing the spirit trade and the use of promissory notes, which had made *immense fortunes* for the officers and some others. Bligh gathered letters from other allies supporting this version of the past. In March William Gore, giving his address as *Cells, Sydney Jail,* wrote to Viscount Castlereagh describing Bligh as the people's *common friend* and Macarthur as *the principal actor in this disgraceful transaction.*

At Government House the committee of rebels found a copy of the comments on Atkins that Bligh had sent to the Secretary for State (*accustomed to inebriety ... the ridicule of the community; sentence of death has been pronounced in moments of intoxication*). These were shown to Atkins, turning him against Bligh so that he later appeared as a witness for Johnston's defence. Bligh's swords were taken away, as were all the firearms in Government House.

Bligh was kept under house arrest, allowed to walk in his garden accompanied by a soldier at a distance of six paces. Some of his supporters were punished. Gore, for instance, was imprisoned for eleven weeks and four days and then sent to work in the coal mines on the Hunter River for four months. On 2 February Macarthur's interrupted trial was continued. It was a farce during which he bullied those witnesses who were Bligh supporters: he was found innocent. One notes that five of the six officers on this occasion had been among those on the bench when he'd refused to accept Atkins' involvement. So presumably he would have been found not guilty on that occasion had the trial continued. On this occasion he was chaired around the town by soldiers to celebrate his acquittal.

The court then turned its attention to the unfortunate George Crossley, who was charged with acting as an attorney after being convicted of perjury. He was sentenced to seven years in the Newcastle coal mines, *to prevent him [Johnston's words] from agitating this settlement again with the contrivance of new villainies.*

On 8 February a public meeting was held in the church to discuss the *ever-memorable* recent events. According to reports received by Bligh, Macarthur distributed spirits and wine to the soldiers and people beforehand, the meeting being held at 8.00pm by candlelight. It was decided to give Johnston a sword and the six officers a present of plate as expressions of gratitude, and to send Macarthur to England to explain the rebellion to the government. (Accepting this task, Macarthur spoke movingly of recent events, frequently clasping his hands together with great energy.) A large amount of money was subscribed for these purposes, the main contributors being Lord, Kable and Underwood, Bayly, the Blaxland brothers, and Blaxcell. Did they recall that at the time of the Glorious Revolution, the London merchants had lent William of Orange £200,000 to help him defeat James II?

The meeting moved votes of thanks to the New South Wales Corps. John Macarthur was also thanked, as *having been chiefly instrumental in bringing about the happy change which took place* on 26 January. Macarthur himself was more modest, acknowledging *the timely interposition of Divine Providence in rescuing him from the malice of his enemies.*

An interesting aspect of this meeting is that it was attended by Isaac Nichols, whom we saw being set up by Macarthur and unfairly convicted in 1799 for obstructing some of the officers' dodgy practices. Since then Nichols had done well, and he now subscribed £50 to support Macarthur and the rebellion. Nichols is an example of how Bligh had managed to offend so many of the colony's wealthier inhabitants, no matter what their relationships with each other.

Four days later, Johnston created the new position of colonial secretary and appointed Macarthur to fill it. The job kept him in Sydney and he soon wrote to his wife: *I have been deeply engaged all this day in contending for the liberties of this unhappy colony, and I am happy to say I have succeeded beyond what I expected.* He urged her to be cheerful: *your headache will then go off – Keep Elizabeth out as much as you can, for exercise in this warm weather is more likely than any other thing to bring about her recovery.*

There was no salary with the secretary's position, but it meant that, where civil matters were concerned, Macarthur now had most of the power of the governor of New South Wales. While he was to favour his allies somewhat, he acted relatively moderately. Many of the rebels were disappointed by this and began to mutter among themselves and complain about the way Macarthur carried out his new duties. There were allegations that he had given away government cattle, but possibly this was done simply because the rebels had no other way of paying the government's bills. Even the trade in spirits was restrained. For the vast majority of the inhabitants of the colony, the change in administration made little difference at all.

One thing Johnston did do was make a large number of

land grants, although all these were subject to approval from London. Johnston gave his own son 2,000 acres.

Their behaviour suggests the rebel leaders were more interested in preparing to justify their action than in reaping any rewards from it. The rebellion had been an ending of something, not a beginning. The events of 26 January reflect not self-interest but self-expression, using the methods available to those involved.

There have been attempts to prove that the rebellion was partly motivated by a desire to reform the system of administration in New South Wales. Alan Atkinson points to a document he believes Macarthur drew up at this time, calling most importantly for a legislative council of four officials and two settlers to advise the governor. He also wanted a public distillery, a system aimed at reforming convicts, and for the colony to have its own London agent. Some of these measures were far-seeing, but as they do not seem to have been mentioned either in the lead-up to the rebellion or in its aftermath, including the efforts to justify Bligh's overthrow, I doubt they were among Macarthur's chief motivations. He doesn't seem to have communicated them to any of the other leading rebels.

Another explanation offered for the rebellion is that it assisted Macarthur's trading activities. In 1807 he had become involved in a complicated trading venture with Robert Campbell and others that concluded with a ship full of goods from Canton reaching Sydney in late 1808. Due to problems along the way, Macarthur's continuing involvement in the deal was in doubt, but as colonial secretary he was able to compel Campbell, the owner of the goods, to sell them to him for a good price. Historian D.R. Hainsworth has suggested that the prospect of the ship's arrival *influenced* Macarthur to overthrow Bligh. It's an interesting idea, but perhaps puts too much weight on what was only one business deal in a life full of them. It also, like other theories previously examined in this book, attributes too much power to Macarthur and not enough to Johnston. However, it's true that Macarthur took advantage of Bligh's overthrow for his trading activities. In February 1808 he sent the *Elizabeth* to Fiji to collect 120 tons of sandalwood, which was then taken to Canton and sold there, something Bligh would never had permitted because it was in breach of the East India Company's monopoly.

One thing we do know is that Johnston brought Michael Dwyer, the Irish man Bligh had transported to Norfolk Island, back to Sydney. This supports the suggestion that Johnston and the other officers had been genuinely outraged by Bligh's action in effectively overturning the decision of the court that had found Dwyer not guilty, and that this had influenced the belief that Bligh might use the magistrates to imprison the six officers after Macarthur's trial was aborted.

Meanwhile, Bligh was assembling his own case and on 30 April wrote a long letter to Viscount Castlereagh describing Macarthur as *the archfiend … whose very breath is sufficient to contaminate a multitude.* The letter contained several important errors. Bligh claimed that Macarthur had attended the corps' dinner on 24 January, and had tried to employ Crossley for his own defence. Later he withdrew the first of these allegations, and never repeated the second, which he would presumably have done had it been true.

Bligh and his allies claimed that the rebels were taking advantage of their opportunities to enrich themselves at the public expense. Certainly some of this occurred, but the rebels were already falling out among themselves, not least because Macarthur refused to give some of the others what they wanted. In late April Johnston sent a letter to the civil and military officers noting the discontent that had *entirely arisen from the confidence I have reposed in Mr Macarthur* and asking them to put any complaints they had in writing. Most of the officers replied in a group letter to the effect that they had no problems with Mr Macarthur. But on 22 May Abbott wrote to Piper on Norfolk Island that *Macarthur's appointment as Colonial Secretary, and the measures pursued, has given almost universal dissatisfaction here, for my own part we disagreed in the early stage of the revolution, if I may term Governor Bligh's suspension so, and don't speak, several others in the same predicament as myself and Bayly has also cut his acquaintance.*

In the same month Macarthur told Piper, who seems to have been hearing about the new regime in stereo, that *some of your old acquaintances have behaved most scurvily – Abbott amongst the worst … Bayly, for whom every proper thing has been done, is become a violent oppositionist … I am of opinion that, had they been given way to, the whole of the public property would not have satisfied them. … In short, if I exempt Kemp, Lawson and Draffin, there is not a man that affords Johnston the least support, and most of them oppose everything, although the whole called upon him to assume the government, and pledged their words of honour to support him.* Indeed, Macarthur had become so unpopular in some circles that he might have risked assassination. Also in April, Gregory Blaxland called in on Anthony Kemp one evening and asked him to come out of the house as he had something serious to relate. Swearing Kemp to keep his identity as a source of this information secret, Blaxland then said that some of the Hawkesbury settlers were planning to kill Macarthur. There were people willing to die in the attempt. When Kemp said this seemed unlikely, Blaxland avowed there were already men from the Hawkesbury in Sydney, and more were on their way. Blaxland admitted he had no love for Macarthur, who had caused trouble for his brother and himself in a commercial dispute, but he *could not bear the idea of murder being committed.* With that he headed off into the night. Kemp decided to make a statement to his colleague, William Minchin, in his capacity as justice of the peace. Minchin noted that Kemp, the deponent, said that *Mr Blaxland appeared to him much confused*, although *from what cause deponent cannot tell.*

In July 1808 the plant collector George Caley wrote about the rebellion to Sir Joseph Banks, his employer, saying *you must have nettled him [Macarthur] when he was in England, from the animosity he bears you. He has circulated a report that you are an old debauched character, and that if you interfere in the present business he will soon cut you down.* Palmer told Bligh later in the year that Macarthur had used his position of colonial secretary to raid the commissariat and put large quantities of his own old mutton in it. Alcohol was given to officers and soldiers (or, in Palmer's words, *to his favourite banditti*). The soldiers also received increased rations of meat. He had given several hundred of the government's cattle to his allies, and had used the government factory to *work up his wool.* Threats of imprisonment and hard labour at Newcastle were used by Macarthur against those who declined to do his bidding.

On 29 July Colonel Joseph Foveaux arrived in Sydney from London, on his way to Norfolk Island. On learning of recent events in Sydney, he took over from Johnston and decided not to reinstate Bligh. In August he wrote to Paterson: *I have already perfectly satisfied myself that Captain Bligh has been acting on a settled plan to destroy and ruin the better class of inhabitants, and that Major Johnston is in possession of incontrovertible proofs of his being guided in the most important concerns of the colony by the advice of Crossley.* Foveaux had left Britain in February, not only after the rebellion but probably after Colonel Paterson, lieutenant-governor of Port Dalrymple and the senior of both Foveaux and Johnston, had learned of it. Yet in the time it had taken Foveaux to sail halfway round the world, Paterson had still not managed to come to Sydney from Van Diemen's Land. (There's a possibility he was misled by Johnston into believing Foveaux would be his superior.) On his arrival, Foveaux asked Paterson to come to Sydney and dispensed with Macarthur's services as colonial secretary.

London, September 1808

The newspapers published various accounts of the rebellion, and Bligh's enemies exulted while his supporters plotted. *The king's authority can never be maintained in that settlement, unless a certain person is removed for ever from the colony*, Samuel Marsden, staying in England, wrote to Sir Joseph Banks. *I need not mention his name. His conduct has been very conspicuous in the late revolution. His great wealth gives him influence in that society; his unbounded ambition makes him turbulent, and his natural ability (which is very considerable) makes him dangerous to the government.* None of which would have been news to Banks.

Edward Macarthur had been sent back to England with despatches from Johnston justifying the rebellion. There he found that Philip King had just died. The Macarthurs' old friend, Watkin Tench, saw a letter Edward had written to his uncle James. Tench, who had been a prisoner of war in France for some time, was now a lieutenant-general of the marines at Plymouth. He wrote to Edward that the letter had aroused in him *the most lively feelings on your dear father's and mother's account*, and assured him that he was *firmly of the opinion that Governor Bligh, by tyranny, oppression and rapacity, has drawn upon himself the just resentment of the inhabitants of the colony, and met with that spirited opposition and final defeat which I trust all unprincipled despots, whether in courts or cottages, always will encounter.* This might not have been inspired solely by friendship with the Macarthurs. In 1795 Tench had written from France to a friend saying: *I still think a limited monarchy the best of governments. Had I been born a Frenchman, I should have struggled as hard for the revolution of 1789, as I should have resisted with all my might that of 1792. Much as I hate despotism, I am scarcely less a foe to democracy.* We don't have such a succinct statement of political principles from Macarthur, but if we did it would doubtless be similar. Bligh had been a despot.

Tench gently advised young Edward of the importance of the rebels not falling out with each other. *The concealment under the feather-bed made me smile*, he wrote, *but did not surprise me in the least, as I had long possessed the strongest testimony from a friend who had served with Governor Bligh that he was not only a tyrant but a poltroon.* He also, helpfully, recommended that Edward be careful whom he talked to in England, as London was full of people who would pass on private conversations to the government. He concluded by saying that whenever he was asked his opinion, as often occurred *at the tables of generals and admirals*, he would proclaim his *perfect conviction of the worth and integrity of your father.*

Edward was staying with W. Thompson, the Macarthurs' London agent, and reported to his mother that his brother John, whom she had last seen when Macarthur took him to England at 1801 at the age of seven, had turned *remarkably stout* and intended to become a lawyer. He reported that *our late affairs make little impression on the public mind*, which was more occupied by the Peninsular War. He delivered Johnston's letters to the Duke of Northumberland. Johnston was relying on his patron to see him through this affair. Northumberland was informed and on side, and gave young Edward two hours of his time, which augured well. Edward also called on George Watson, Macarthur's old ally in persuading Lord Camden to give him the Cowpastures. These friends, together with James

Macarthur, helped Edward counter a *most virulent* article published by Bligh's allies in October 1808. Edward believed Samuel Marsden, then in England, had had a hand in the publication, and told his father: *indeed he does not scruple to say that the business is at issue between himself and you, and that one or the other must desert the country of New South Wales.*

Edward also reported that General Grose was attentive, as was Henry Brogden, who had been helped financially by Macarthur and whose brother, James, was a member of parliament. Edward was received by Hugh Elliot, brother of the governor-general of India and a powerful Whig. He was welcomed *as usual* by the family of Sir Walter Farquhar, the Macarthurs' main patron, but Sir Walter *does not at present take an active part, although at first he talked a great deal.* The amount of time and effort devoted to lobbying and rallying support was considerable, and the class of person the young man was able to meet reflects his father's impressive connections in Britain.

However, the Macarthurs' friends were all worried that, while Bligh had many enemies, the government was greatly concerned about the precedent that would be established should the rebels not be punished. And the Tory Lord Castlereagh, Secretary of State, who had helped suppress the 1798 uprising in Ireland, sympathised with Bligh. Edward wrote home: *The two emus arrived safe and were presented to Lady Castlereagh, and one swan and a goose lived, which were given to Lady Camden. Mr Watson desired me to say that their ladyships were desirous of having some bronzewing pigeons; but he would not permit me to give the pair I had, because he thought that so many presents at one time would overdo the business.*

Edward also reported on the condition of some wool that had come over in the *Dart*, which an expert had said would make *very good cloth*. Then, having done an excellent few weeks' work, the busy youth purchased a commission in the army and went off to Spain to fight the French and begin a successful military career.

Sydney, 1809: The Third Duel

Nothing is known about Macarthur's activities for the rest of 1808. On 9 January the following year, Colonel William Paterson finally arrived from Port Dalrymple and assumed command at Sydney. Ten days later, Macarthur fought a duel with Foveaux. It was his third duel. He was forty-two and Foveaux was forty-four. Foveaux believed Macarthur had not accounted for £500 in the accounts he'd handed over when relieved of his position as colonial secretary. He had asked Macarthur to repay the money, and in reply was challenged to a duel. This might not have been about money – according to James Finucane, Foveaux's secretary, Macarthur was still upset at the loss of the colonial secretary's position.

Initially the challenge was refused. But once Paterson arrived, Macarthur again challenged Foveaux, who this time accepted. There is a strange letter from Macarthur to Johnston, who was his second, in which he encloses a bill and, *as nothing but money will induce the gallant colonel to take the field*, asks the major to give it to Foveaux to *pay my obligation for 500 pounds, and I will thank you to get Lieutenant Colonel Foveaux's receipt*. He then added: *I propose going to Parramatta tomorrow, and shall be prepared to join you on the road towards your house any time the next day you may determine*. The duel was fought near Johnston's house in Annandale. Finucane, who was Foveaux's second, recorded that a coin was tossed for the privilege of shooting first. Macarthur won and stood ten paces from his target, who was quite fat: *He took a very deliberate aim and was perfectly cool, yet missed his object which was one of no small magnitude.*

Foveaux declined to return fire, saying he had no idea what the duel was all about. However, he did suggest it was hard to have to *account in that manner* for something he'd said as part of his duty as acting governor. Johnston and Finucane now persuaded Macarthur to end the matter without any more shooting, which he agreed to do. Finucane noted that *this disagreeable affair ended in a reconciliation which I trust may prove cordial and lasting.*

Macarthur turned his attention to his forthcoming departure for England, where Johnston and he were to attempt to put their case to the government, and advertised some of his stock and 2,000 acres at Seven Hills on the Hawkesbury road, for ready money only. The ads ran in the *Sydney Gazette* for many weeks but there was no buyer, an interesting indication of just how non-portable the assets of the New South Wales gentry were.

Like Foveaux, Paterson declined to reinstate Bligh. He wrote that Bligh *bore the most rancorous ill-will to every officer and inhabitant who he conceived could possibly in the remotest manner interfere with a matured plan of exercising the high command with which he was honoured in the purposes of gratifying his insatiable tyrannic disposition and advancing his pecuniary interest.* Paterson was keen on the quiet life and must have regarded the situation with considerable distaste. We don't know if the ball from Macarthur's pistol, acquired in their duel in 1801, was still lodged in his shoulder. But he was not well and now proceeded to spend much of his time out of the way at Government House at Parramatta.

Paterson's problem was to get Johnston and Bligh back to England, where the government could decide what it wanted to do. Johnston was planning to leave on the *Admiral Gambier*. At first there was some talk of Bligh sailing on the same vessel, but naturally the idea was repugnant to him.

Eventually Lieutenant Finucane brokered an arrangement whereby Bligh and his daughter would be allowed to return home on the *Porpoise*. Finucane drew up a contract to this effect, and multiple copies were signed by Bligh and Paterson. Bligh and his daughter were allowed to board the ship, but he promptly broke his promise and sailed down to Hobart to seek assistance from Colonel David Collins, who had abandoned Port Phillip and moved his small settlement to Van Diemen's Land. Johnston later commented that Bligh's breaking his pledge in this manner might be vindicated, *but I confess I neither understand, nor wish to understand, the code of honour from which such a vindication must be drawn*. At first Collins was friendly, but soon he turned against Bligh. He was not even a member of the New South Wales Corps, yet he too apparently found Bligh's record as governor repugnant. Exposed to Bligh's personality, he commented: *God knows I never had any malice in my heart until I came into contact with this detestable brute*. Bligh and his daughter remained on the *Porpoise*, moored in the Derwent and with no supplies from Collins, for the next year.

Johnston was keen to take with him to London a number of witnesses who could help demonstrate the justice of his actions to his superiors. The rebel leaders departed Port Jackson in the *Admiral Gambier* on 29 March 1809. There was a big party at Government House several days earlier, and the *Venus*, a small ship loaded with well-wishers, sailed forty miles out to sea to see them away. Macarthur had his young sons James and William with him, off to obtain an education as their older brothers had done before. They were leaving behind with their mother their sisters, Elizabeth the invalid, Mary, and little Emmeline, who had been born the year before and was the last of the Macarthurs' seven surviving children.

John was going back to Britain to face possible censure and disgrace. It was a grim journey for the foremost settler in

New South Wales, yet he could hardly have expected anything else when he overthrew Governor Bligh.

★

Conflict had played an important part in Macarthur's life, not just the duels, with Gilbert back in Plymouth and with Paterson and Foveaux later, but the close calls with Balmain and Marshall and the refusal to fight Hunter. As we have seen, King had refrained from calling Macarthur out only due to his sense of the responsibilities of his position. There was nothing honourable about Macarthur's behaviour in several of these clashes, even though they were conducted according to something called the code of honour. The duel, of course, was not honourable by our standards – it could redeem the guilty and kill the innocent. Still, the code survived because it served a purpose. It filled the vacuum when the rule of law was weak or irrelevant. In the early 1800s the code of honour was struggling with the law, and slowly being strangled by it. Standards of morality were confused, and sometimes – as Bligh had found – you might find yourself using the law when your opponent had switched to the code, or, as Lieutenant Marshall discovered, vice versa. But the code of honour helps explain both the January 1808 uprising and the response of the British government and army.

Apart from the wealth of circumstantial evidence, proof of the existence and strength of the code of honour among the gentlemen of New South Wales is to be found in their language. As I have shown, they referred constantly to their honour and their status as gentlemen. In the early twenty-first century it can be tempting to trivialise and dismiss such talk as at best quaint, at worst self-serving rhetoric. In fact the concepts expressed by these words sometimes guided their speakers' lives.

John Macarthur was particularly suited by background and character for such a situation of moral complexity. He was Australia's first great bush lawyer, capable of using his keen mind to push the legal system as far as it could take him. But he also had a capacity for violence that enabled him to make full use of the duel – or the threat of a duel – where necessary. And he had the rhetorical skill and the psychological complexity needed to switch between and even combine these two modes of conflict. Even by the standards of his time, he was an extreme character.

*

Macarthur reached England with Johnston in November 1809. They hoped that the government would soon let them know what it thought of their actions, and Johnston demanded a court martial, but nothing occurred for over a year because Bligh refused to leave New South Wales until the new governor, Lachlan Macquarie, arrived. This duly occurred and Bligh left Sydney, along with the New South Wales Corps, which the British government, its interest in the colony briefly awoken by the rebellion, had decided to withdraw.

Macquarie had enough time with his predecessor to reach an opinion of his trustworthiness – of his honour – with which Foveaux and others would have concurred. *It is impossible*, he wrote, *to place the smallest reliance on the fulfilment of any engagement he enters into*. More generally, he observed: *Governor Bligh certainly is a most disagreeable person to have any dealings, or public business to transact with … his natural temper is uncommonly harsh, and tyrannical in the extreme.*

Waiting in London, Macarthur's spirits swung between

hope and despair. The letters he wrote to his wife in these years are almost the only personal record we have of his thoughts and feelings. One gets the sense when coming upon them that one is uncovering the real Macarthur at last; most of the previous records of his life deal only with his public disputes. Yet, while it is refreshing to uncover his feelings about his family and his health, in these letters we are seeing him in an unusually vulnerable position, sick, lonely and increasingly broke. The enormous wealth that was to come to the Macarthurs from wool was still a long way off, and in this period they lost a great deal in failed trading ventures. A depression in the colony reduced their income from the land. So for many reasons he was not himself.

The letters are full of details about his children. Elizabeth junior was so ill that for a while her father expected to hear any day of her death. Her health varied, and in one letter he learned that she had lost the use of her limbs. Of the boys, Edward was an ensign in the army in Europe, John junior was enrolled as a law student at the University of Glasgow, and James and William were in school.

John Blaxland had returned to England already and been gaoled, at Bligh's request, in Cape Town and sent to Britain as a prisoner. This did not augur well for the rebels. On the other hand, the government had changed, which was good – Macarthur believed the outgoing administration might have *crushed us altogether.* But as he acknowledged, with the war against Napoleon raging, no one in Britain had much thought for New South Wales: *No man thinks of another's safety when his own house is burning.*

Macarthur found the delay frustrating: *to live in such a state of suspense is dreadful.* If he had *some employment to occupy my time my situation would be less painful, but I have not the slightest occupation.* He became ill, soon after his arrival being *seized with violent spasms in my side which increased rather than abated in their violence*, for several days. He diagnosed this as *a most violent cold*, which after subsiding for a few days *returned again with renewed violence, and in a very short time had so increased as to confine me to my bed. The complaint continued with but little dimunition for several weeks.* He sought relief at mineral springs and by changing his address in search of better air, but without success, in May 1810 describing his mysterious illness as *a terrible nervous affection.* Although he was being treated by the Prince of Wales' own doctor, he did not improve. He also suffered from a *complaint on my lungs*, and *continual slight attacks of the liver accompanied by the old plague obstinate indigestion.* For six months he drank nothing but water and some milk – no tea or alcohol. He ate nothing but bread, fruit and vegetables. *For my breakfast I eat bread and milk my dinner consists of bread potatoes and sometimes ripe currants or gooseberries. … I take a great deal of exercise, and do it without feeling fatigue, whereas whilst I eat animal food … a walk of two miles was the utmost I could perform.* Behind this concern with all his illnesses we sense his dreadful frustration as he waited, bored, relatively powerless and afraid, for the government to act.

Macarthur spent time talking to his lawyers and friends and arranging political support. Bligh had included Walter Davidson in his proclamation against the rebels, which naturally upset his uncle, Sir Walter Farquhar. As we've seen, Johnston enlisted the aid of the Duke of Northumberland. On the other side, Sir Joseph Banks and Samuel Marsden, who was still in England, were agitating for the absent Bligh. They had so much success in spreading stories against Macarthur that he complained *I think the people of England the greatest dupes in the universe.* He counterattacked, and was able to claim at one point that *Bligh is now universally execrated, before we arrived he was pitied.* In his many letters he never showed a sign of regret for what had happened, even as the number of months increased.

The situation was complicated by another Madras mutiny, which occurred around this time. Governor Sir George Barlow was deposed. Some thought it was partly inspired by the Bligh rebellion, which would have had members of the government scratching their heads and thinking about *precedent.* It was big news – far bigger than the events in New South Wales – although its effect on Johnston's trial is unknown.

Macarthur's letters were full of his thoughts on his relationships with other people, especially ones important to him. Always there is a movement towards those higher than himself and away from those lower. Indeed, he even, somewhat ungratefully one might think, told Elizabeth that it was good for the colony that the corps was being withdrawn, for *a more improper set of men could not be collected together than they have lately become.* In particular he identified his duelling opponent, Foveaux, as *that unprincipled man.* Presumably his experiences with the officers during his period as colonial secretary had soured his relationships. Yet Johnston and he were relying on them for support.

Macarthur still played his games. His old foe William Paterson had died on the voyage back to England, and when his widow arrived at Portsmouth in November 1810, Macarthur met her and *returned to town yesterday bringing under my escort Mrs Paterson, who appears to be grateful for this mark of attention – you know I sometimes like to return disobliging acts this way.* When Foveaux arrived in England, Macarthur acted towards him too in a friendly manner. He enjoyed surprising people, keeping them on edge. *Foveaux appeared as much pleased as astonished*, he wrote to Elizabeth. *When shall I be known?*

When indeed?

*

Bligh reached London in October 1810 and Johnston's court martial began the following May at Chelsea Hospital. The cast was impressive: twenty-two witnesses for the prosecution and eighteen for the defence, many brought from New South Wales just for the trial. They included John Palmer, Robert Campbell, William Gore, Francis Oakes, Nicholas Divine, Richard Atkins, Anthony Fenn Kemp, John Harris and Macarthur himself. The trial ran for thirteen days and we are tremendously fortunate that Bligh paid someone to record it in shorthand, publishing the transcript in November 1811. Johnston was charged with mutiny and the judges were the judge advocate general, Charles Manners Sutton, and fourteen senior soldiers headed by Lieutenant-General Keppel. Both Bligh and Johnston were represented by top lawyers.

What does the trial tell us about the causes of the Bligh rebellion? Almost nothing we do not already know. There were no new arguments, and the court was vague about the criteria it used to arrive at its eventual decision, for which it gave almost no reason. Witnesses were often allowed to contradict each other and even themselves with no comment from the court. In his opening speech, Bligh said he was *at a loss to conceive* any justification for the mutiny. But he went on to say Macarthur had been dissatisfied when the barter of spirits was stopped. Later he claimed that the others involved in the mutiny – including the officers – had also been involved in this barter. Yet Bligh's witness John Palmer disagreed, and said *I knew some officers who had an interest in the sale of considerable quantities of spirits and wine; but I cannot say that was the cause of the mutiny; and it was a long time prior to the 26th.*

Johnston, who told the court he had been the first officer

ashore when Sydney was founded in 1788, said he'd deposed Bligh only because of his desire to prevent a massacre and the plunder and ruin of an infant colony. He had been afraid of what would happen should the six officers be imprisoned. Historians have tended to downplay this argument – as did Bligh and his witnesses during the trial – but it is worth repeating that rebellion by the mob was a constant fear during this period. (In May 1810 Macarthur, writing to his wife from London, had told her *Many are of opinion that a revolution is unavoidable.*) The court did take evidence from various witnesses on whether Johnston's fear was realistic, but the results were conflicting and inconclusive.

In his final speech, Bligh sought to sideline the very crux of Johnston's defence, claiming that he had merely asked the six officers to appear before him *with an intention of pointing out the illegality of their conduct, and hoping to bring them to a sense of their duty*. But, he assured the court, he had no intention of arresting them. Johnston of course said he believed the officers would have been arrested unless they backed down, and there seems no reason to doubt that Bligh's rhetoric of treason gave him good grounds for this belief. He told the court: *Their conduct had no affinity whatever to treason or treasonable practices; and the attempt to punish or proceed against them, on that ground, was indeed most flagrant and most alarming.*

And yet, what if William Bligh was not being disingenuous, but was telling the truth when he said he had not intended to arrest the officers? To have done so would have been reckless, and he was not a fool. But he was an incompetent communicator. So it is possible the rebellion occurred because of a mistaken impression on Johnston's part, caused by Bligh's use of the wrong words, words which no gentleman should have used about other gentlemen. But where honour was concerned, Bligh had the equivalent of a tin ear. It seems that here, as before, Bligh had used language extremely loosely, without realising the effect it would have on a certain type of audience.

The issue of Bligh's threats to leases was also thrashed out in court, also inconclusively. Bligh said he had asked some leaseholders to move, but never got to the point of ordering them to do so as he had written to the secretary of state asking for authority to do this, and was still awaiting a reply in January 1808. *I never did deprive them*, he said, ... *I never deprived any one man of his house, land or property, in my life.* And yet later he acknowledged that two people did move from the area behind Government House and he had ordered Macarthur to move. John Palmer and William Gore, witnesses for Bligh, said that some people were moved from their houses. But hardly any. So once again it appears that Bligh had managed to evoke a response in those he dealt with out of all proportion to the actions involved. And once again, wild and offensive language had been central to this.

Other issues were canvassed during the trial, such as Bligh's suppression of promissory notes written in colonial currency, and his putting a stop to the illegal harsh punishment of convict servants. But they were not substantial. The rebels did go to considerable lengths to prove that Bligh was a coward for hiding under the bed, and the court allowed plenty of evidence to be presented on this point. Before the trial, Johnston had assured the government, in a letter possibly written by Macarthur, that he would prove that Bligh had *been guilty of heretofore unheard of and disgraceful cowardice*. He clearly believed that the mutiny was intricately linked to Bligh's character as a gentleman, and that other people would view it in the same way. Therefore he was determined to destroy that character. During the trial he said Bligh had been found in Government House in *a condition which would make the real heroes of the British navy blush with shame and boil with indignation.* To which Bligh later responded, *I certainly at such a suggestion burn with indignation, but who ought to blush with shame I leave others to determine.*

The statements to the court suggest that William Bligh was deposed because he came to be seen, by his peers in New South Wales, as a man without honour, so bad a ruler that it was impossible even to negotiate with him. The gentry of New South Wales were more primitive in their ideas of honour than the gentry in general back in Britain, partly because the law was so weak in the colony and partly because of the type of men they were, most of them soldiers and parvenu gentlemen uncertain about their status. Bligh was in one sense a more modern person, interested in doing the right thing by his subjects rather than just maintaining the status quo. His opponents reached back to a British tradition which gave them the right to expect otherwise. Writing about the two centuries leading up to 1642, historian Mervyn James says that *Honour … implied for those that professed it a consistency of public and political attitude which was summed up in 'faithfulness' to lords and friends. It constituted an appropriate code of political conduct for the kind of political conflict in which lords confronted unworthy kings, and which in the last resort was resolved by the violence of war or battle.* In Tudor England, according to James, a civil society had begun to emerge *in which the monopoly both of honour and violence by the state was asserted. Equally important in bringing the change about is the more or less coterminous process which I shall call 'the moralisation of politics'.* This process collapsed under the Stuart kings, but it reads like a dry run for what was to happen during Macarthur's lifetime, when William Wilberforce and his allies were proclaiming the need for the state to involve itself much more in moral matters, including the monopolisation of violence between individuals.

This is one reason why Johnston was limited in how much

he could talk about honour during his trial: he was being watched by a rising middle class whose members rejected much of what men of honour stood for. He had to get the message across to his judges – who, as soldiers, knew about honour and life on the empire's frontiers – without upsetting the lawyers and other civilians who had never held a pistol or left Europe. Johnston, and Macarthur, were caught in the death struggle between the code of honour and the rule of law. Whatever decisions they had made in New South Wales, once they returned to Britain they found they had placed themselves on the losing side.

But they had friends. The morning after the taking of evidence concluded, a newspaper called *The British Express* published an account of the sentence on Johnston, including comments by the court that were damningly critical of Bligh. A copy of the newspaper, the ink still wet, was sent immediately to Bligh in an envelope with a pistol drawn on its inside, the implication being that he should kill himself from shame. The entire thing was an invention.

In reality, Johnston was found guilty of mutiny and cashiered, but allowed to return to his house and his family in Sydney. It was a good result – usually an officer might be cashiered for a minor offence, such as being drunk on duty. Perhaps it was not as good as Johnston had hoped for, but it was much better than he might have feared. Before the trial, as informed an observer as Lachlan Macquarie had believed Johnston would be executed for the rebellion. On their way to New South Wales, the Macquaries had met Thomas Jamison and John Harris at Rio de Janeiro and heard all about the rebellion from these witnesses. Elizabeth Macquarie wrote in her diary: *We felt sorry that a man such as Colonel Johnston was described to us, should have committed himself as he has done, by an act of the most open and daring rebellion, by which in as far as it appears to us, he will probably forfeit a life.*

In the end, Johnston was saved because the court was convinced that Bligh had been a bad governor. A brief public statement was published in the *Times* over the commander-in-chief's name: *I am further commanded to acquaint you, that in passing a sentence so inadequate to the enormity of the crime of which the prisoner has been found guilty, the court have apparently been actuated by a consideration for the circumstances of impropriety and oppression which, by the evidence on the face of the proceedings, appear to have so strongly marked the conduct of Governor Bligh.* William Bligh complained and the statement was slightly modified in the newspaper a few days later. But that is what the court had found.

The only other action the government took, as already noted, was to order Macquarie to put Macarthur on trial when he returned to Sydney. Macarthur did not become aware of this for some time – the government might have kept it quiet in the hope he would return to New South Wales. But even before he knew of it he showed no enthusiasm for returning to his family and the colony. Immediately after Johnston's trial he sank into a depression, and then there was the physical illness that lasted for some years. Possibly he felt his health would not stand up to the journey home. Macarthur was to be away from his wife and daughters for a wrenching total of over eight years. He had brought about this strange rebellion and now, perhaps even more than William Bligh, he was to be its chief victim.

⋆

I have sought to show that self-interest, at least of the material kind, cannot fully explain the Bligh rebellion. It fails to explain why canny and experienced men would have risked death when the threat to their material interests was real but not overwhelming, and there was no reason to believe the rebellion would bring a governor any more sympathetic to those interests. It's worth noting that the rebels and many of their contemporaries did not talk of the rebellion in material terms. As noted before, the phrase 'Rum Rebellion' was invented many years later, by a Quaker author of the 1850s. The rebels spoke of it in terms of property and freedom, honour and tyranny. But to succeeding generations taught to analyse history by following money trails, this has seemed like so much self-serving hokum.

The other explanation used by some historians is that Macarthur caused the rebellion through his amazing powers of manipulation. While this was a factor, like the material argument it hardly explains why he was able to persuade the officers – many of whom were indifferent to both his charms and his threats – to act in a manner so contrary to their own interests. It's true his arguments and scheming worked, but only because he was able to mobilise some specific and powerful cultural beliefs.

Perhaps sensing that they do not understand the Rum Rebellion, some observers have sought to trivialise it, bringing in terms such as *farce* and *comic opera*. There's no doubt that a rebellion in which no one was injured and the governor was dragged from under a bed is open to ridicule, particularly when compared with the then recent American and French revolutions. But the Bligh rebellion did not strike its contemporaries as particularly funny. Robert Darnton has written that when in history *we cannot get* a joke *we know we are on to something*. In this case the problem is that we cannot get why something was *not* a joke. Perhaps the meaning of the event lies behind the terms that have been used to denigrate it. *Farce*, in the form of melodrama, and at least some *comic opera*, drew much of their humour in the nineteenth century from their denigration of outmoded, aristocratic values and manners. The Bligh rebellion involved a similar set of values and manners. Macarthur looked backwards, to the old aristocratic ideals filtered through the army, and an ageing obsession with the idea of limited government, while Bligh looked forwards, to the just-emerging ideal of an objective, professional administrator, a position that would henceforth be more powerful than in the past. Macarthur believed, like the philosopher David Hume, that *Reason is, and ought only to be the slave of the passions, and can never pretend to any other office than to serve and obey them.* That was the old basis of government. Bligh, like Bentham and Wilberforce, believed that reason – informed by ideas of good and evil – should dominate the passions. But this most passionate of men ignored the passion of others at his peril.

What are we to make of John Macarthur's life more generally? While he was notorious for the number and vehemence of his disputes, his other activities stood out only because of their scale. The officers have been often criticised for their monopoly trading practices, but some of these criticisms are misplaced, not least because the practices represented behaviour common at the time. Macarthur was more successful than most of his contemporaries who engaged in the same activities, and therefore more prominent. He operated generally within the law, and it was not his fault if the law in his time was flexible, at least for gentlemen.

There was a benefit to the public in the activities he undertook. As Adam Smith noted when Macarthur was a boy, business people acting greedily – albeit regulated by government – are promoting the general prosperity. Macarthur created jobs, helped build industries, and drew money into Australia. If we compare his activities not with their equivalents in well-regulated twenty-first-century Australia but with New South Wales as it would have been had he and others not built their businesses, which is the only realistic alternative, we see a slothful peasant society in which people might have starved and the more energetic officers, lacking trade as an outlet for their energies, would have turned to more violent and socially harmful activities. So, if one believes that commercial activity is a civilising force and that property ownership is the basis of democracy, John Macarthur can be seen as both trader and hero. And if one believes that the first step on the road from autocracy to democracy is for someone to defy the autocrat, then John Macarthur, whatever his motives, was on the side of freedom.

Epilogue

By mail Macarthur sought eagerly for information about Governor Macquarie and his activities in Sydney. At first he had high hopes, looking forward to *the changes such a man will make amongst the wretches who have so long insulted every honourable and virtuous feeling by the unblushing display and avowal of infamy and vice.* Macquarie visited Elizabeth Macarthur four days after his arrival in Sydney, and wrote to London approving Foveaux's decision not to reinstate Bligh. It seemed that Macquarie was a man whose judgement might be relied on.

But in April 1811 Macarthur wrote to his wife exhibiting an almost hysterical degree of concern at developments in New South Wales. Referring to her most recent letter, he cried: *Every paragraph increases my amazement, and every circumstance you relate, adds to the perplexity of my mind. – God alone knows how such a state of things as you describe may terminate, or how operate upon our affairs – Would to God I could withdraw you all from the colony but it is fruitless to indulge wishes, or even to form plans of future proceedings, whilst we are hurried forward by a course of events, that seem to set human prudence and all ordinary calculation at defiance.*

The news was out in England and Macarthur was being queried by everyone about the terrible changes in the colony. *Is it possible, it is said, that Governor Macquarie can associate with, and bring to his table, men who have been convicts? Who have amassed fortunes by the most infamous frauds, and have and continue to set the most shameful examples of dissoluteness and vice?* Such developments, he wrote, *fill my heart with dismay and grief. ... In truth I know not how to act, or what to say, and the more I think, the more distracted and puzzled do I feel.*

John Macarthur had dragged himself up from modest beginnings to the honourable status of a gentleman. It meant more to him, perhaps, than to many men who had been born at that level of society. The social group he had entered linked him, however tenuously, to the aristocrats and the king, and all the other rulers of England, past and present. He had striven to live by his somewhat romantic notion of the code of honour, which had included fighting three duels and deposing a governor. All this for him was tangible, as real and important as money, sex or food, as life itself. It had enabled him to survive, and prosper, in a frontier colony full of evil men and evil deeds. And now Lachlan Macquarie, by treating emancipists the same as gentlemen, threatened to make Macarthur's great achievement, his concept of himself and his society, meaningless. This blow hit him with the force of something deadly to his spirit. He knew not what to say, or how to act.

In Macquarie's actions he had glimpsed an Australian future very different from the one he had hoped for. Yet with the Bligh rebellion, he himself had helped bring it into being.

Appendix

Some Letters and a Speech

Two Letters

These two letters show the beginning of the long-running dispute between John Macarthur and Richard Atkins, and also give an idea of the bitter nature of many exchanges of the time. In July 1796 magistrate Atkins refused to tell Captain Macarthur the name of a soldier caught stealing turnips from the garden of Government House, Parramatta. Macarthur responded with a letter insultingly addressed to *Mr* Richard Atkins rather than Richard Atkins *Esq*. When Governor John Hunter reproved Macarthur for this, he received the following letter in response.

CAPTAIN MACARTHUR TO GOVERNOR HUNTER.

Sir, Parramatta, 25th July, 1796

I had the honour to receive your Excellency's letter of the 23rd instant last evening, and I confess my astonishment at the manner in which you have been pleased to answer my complaint against Mr. Atkins.

I complained of having received a gross and wanton insult in the execution of my duty; but instead of receiving redress I am told my having omitted to bestow the title of Esquire on Mr. Atkins was intended to wound his feelings as a gentleman, or to lessen him in the eye of the public as a magistrate.

Surely your Excellency cannot seriously suppose that I cou'd design to effect either of those purposes by such insignificant means; for is it possible that the use of a trifling appellation can produce a change in the public opinion of a man so deeply plunged in infamy; or how can he be imagined to possess one feeling of the gentleman when the enormities he is committing daily are considered.

I have complained of Mr. Atkins because I know myself to be injured, and

I persist in my complaint because I think it degrading to his Majesty's service that an officer bearing his Sovereign's Commission shou'd be insulted when in the performance of his duty by a man so vile as is the person in question. If your Excellency shou'd see it fit to call on me for a specification of the facts on which what I have written is founded, and to substantiate those facts by evidence, I shall with great readiness and pleasure come forward. Both by oral and written evidence will I prove that Mr. Atkins is a public cheater, living in the most boundless dissipation, without any visible means of maintaining it than by imposture on unwary strangers whose business leads them to this settlement.

I will prove that in his public and official capacity drunkenness and indecency are almost inseparable from him; and that no longer since than the 19th instant he was exposing himself at an early hour of the morning in the public streets in the most disgracing state of intoxication. I will prove that very recently he in a fit of drunkenness wrote such a letter on business to an officer of my corps as he has since thought it prudent to request might be destroyed, least it shou'd reach your eye, and discover to your Excellency how exemplary a character the distribution of public justice is entrusted to.

I have, &c.,

JOHN McARTHUR.

Atkins responded with the following.

RICHARD ATKINS TO CAPTAIN MACARTHUR.

Sir, August, 1796

The repugnance I feel at descending to an epistolary or any other altercation with a man of your principles can only be surmounted by the gratification that naturally results to a man of honour in delineating even by paper the deformity of an opposite character; and where can I find a subject like yourself?–A man, the baseness of whose heart even imagination, however warm, can hardly portray. To enter into contact with such a being might become contagious; but, however, on this occasion I will not decline giving my unequivocal sentiments of so worthless a member of society.

That your charges, or, as you now term them, assertions, against me were founded upon the most ignoble motives, such as malice, revenge, &c., is incontestably proved by your meanly abandoning them, after positively pledging yourself to support them by "proof oral and written." What must your sense of shame be when you, a Goliah of honour and veracity, should resort to a subtifuge at which the meanest convict might blush, by skulking from substantial meaning and screening yourself by a jingle of words from that manly perseverance which should mark the character of a man professing as you do. The quibble between charges and assertions is of too flimsy a texture to require a comment. It is only worthy of a dastardly coward like yourself. Your original meanness and despicable littleness pervades your every action. It shows the cloven foot. Return to your original nothing; we know what you have been and what you now are; and believe me an honest and industrious staymaker is a more honourable and more useful member of society than such a man as I hold you to be.

Let me ask who has been the incendiary – who has been the promoter of all the feuds and animosities between individuals in this colony? You, sir. You are likewise the man who has had the audacity to accuse me with having acted officially and individually with injustice, oppression, and peculation – nay, even highway robbery. You who, four years ago, was only a lieutenant, pennyless but by his pay, and is now reputed worth £8,000. Let this colony bear witness where lies the strongest presumption, you or me, being the oppressor, peculator, or robber. On this subject, viper, you bite a file; the day of retribution will come, and believe me it is not far off, when you will be dragged forward by the strong arm of justice to public view as a monster of society, the betrayer of private conversation, to answer your own malicious ends, the assassin of all that constitutes true honour. But why adopt that epithet when addressing a man to whom it is a perfect stranger other than by sound, for he never felt it?

I at this period feel myself degraded by devoting a moment in becoming monitor of such a man. Could you suppose that under any possible point of view your pomposity in promising to support your charges against me must not recoil more forcibly on yourself, and render you an object of real contempt with every man of real honour and veracity? Can you believe that the man who has been guilty of such "enormities as you are daily practising" can but be detested by all mankind who have a spark of benevolence and phylanthropy in their composition? No, you are known too well not to suppose it! You have passed the Rubicon of true dishonour; you, however, are seared against its sense, and the less pitiable, feeling it not that you are a leper in reputation, and that you ought to be driven from the society of all good men least you should be infectious. Shall I go on, or have I said enough to a man who bears "his Sovereign's Commission" and has been called lyer and scoundrel, and, if I am not misinformed, received a blow, and still talks of honour? I assert you have been called these; I apply them again to you, and if your fertile imagination can form any epithets more forcible, take them, for they are your own. The manners of a gentleman only causes my subscribing myself,

Your humble serv't,

RICHARD ATKINS.

I think it proper to acqt. that your late infamous transaction respecting me, as well as some other well-authenticated facts, are going home for ye perusal of those whom it may concern; it is not your signature of a few letters that will wipe them away.

(These letters can be found in HRNSW vol. 3 pp. 122ff.)

Macarthur's 'men of honour' speech

The following speech provides an example of Macarthur's use of the court system for political purposes. The occasion was the crucial court case on 25 January 1808, the day before the Bligh rebellion. Following incidents surrounding the seizure of two stills belonging to him, and a quarrel with naval officer Robert Campbell over the ship *Parramatta*, Macarthur had been charged with sedition. In the indictment drawn up for Judge Advocate Richard Atkins by ex-convict and lawyer George Crossley, Macarthur was charged with being *a malicious and seditious man, and of a depraved mind and wicked and diabolical disposition, [who] had been deceitfully, wickedly and maliciously contriving and abetting against William Bligh, Esq., His majesty's Governor-in-Chief of this territory, to vilify and represent him and others as unjust officers and ministers and little fit to be used and entrusted by our said Lord the King with the weighty affairs of this territory.*

To the Members of the Criminal Court.

Gentlemen,

It will, I am convinced, excite your surprise, as I think it must of every impartial man, to hear that I am a prisoner to this bar, utterly unacquainted, except from rumour, of the nature of the accusation against which I am to defend myself. Such, however is a fact, for although I have made three written applications to the judge Advocate, for a copy of the indictment or information, I have not been able to obtain it.

In this unprecedented situation, and having been informed that the charge against me had been founded on certain events which originated in the illegal and arbitrary conduct of the Judge Advocate, as exemplified in the correspondence and warrants, I did conceive it prudent and a piece of duty I owed to the community, to protest against Richard Atkins Esq., being appointed to sit as a Judge on a trial where he is so much interested, and in which his own security is so materially involved.

To prevent unnecessary delay, and other consequences which I apprehended, I did, in a letter to his Excellency Gov. Bligh protest against the Judge Advocate, and respectfully required that a disinterested person might be appointed to preside at my trial.

To this His Excellency was pleased to answer "that the law must take its course, as he does not feel himself justified to use any interference with the executive power by which I suppose it (sic) meant the judicial authority, and I humbly conceive His Excellencys power must be the Executive.

Defeated in this attempt to obtain what I know to be my lawful right, my only alternative is to resort to the Members of this Court, and I do so under an entire confidence that what I can prove to be my right, you as men of honour will grant me.

To you then gentlemen I appeal and solemnly protest against Richard Atkins Esq., being allowed to take his seat as one of my judges on this trial.

To support this protest, my first objection is, because there is a suit pending between us, for the recovery of a sum of money that he unjustly withholds, and, as he is screened from the operation of the law, is to be submitted to His Majestys Ministers.

My second objection is, because I can prove he has for many years cherished a rancorous inveteracy against me, which has displayed itself in the propagation of malignant falsehoods, and every act of injustice that can be expected to proceed from a person armed with power, against a man whose life and conduct is, I trust, a public satire on his own.

My third objection is, because I have long been the object of his vindictive malice, in consequence of my having been called as an evidence to support an accusation made against him by John Harris Esq., that he was a swindler.

My fourth objection is, because he has associated and combined with that well-known dismembered limb of the law, George Crossley, and others of as wicked minds, though perhaps not quite so notorious, to accomplish my destruction.

In proof of this I have evidence to prove that Crosley has prepared the information to be produced at this trial, and has arranged the whole plan of evidence, he being considered eminently qualified to conduct that part of the business from his extensive practice in that particular branch of legal knowledge.

I have also proofs in my hands in the writing of that veteran practitioner Crosley, which will convince the most sceptical mind that other schemes have been agitated to deprive me of my property, liberty, honour and life.

Here it is gentlemen, read it, and after, read the proceedings of a Bench of Magistrates, and you will see that for presuming to complain of a most unlawful seizure of my property, which the Judge Advocate joined in reprobating, it has been determined to ruin me.

This precious document came into my hands as if by the interposition of Divine Providence, it was dropped from the pocket of Crosley and brought to me.

That you may consider it at your leisure, I annex a copy both of it, and of the proceedings of the Bench of Magistrates.

My fifth objection is because Richard Atkins Esq is my prosecutor on this trial, and is so deeply interested to procure my conviction, that, should he fail, nothing but the arm of power can save him from a criminal prosecution, at this very Bar, for false imprisonment of me.

My sixth and last objection is on his having already pronounced sentence of condemnation against me, as is presumptively proved, and can be clearly, by his declaring that the Bench of Magistrates had the power to punish me by fine and imprisonment thereby clearly demonstrating an intention to deprive me of the benefit of my present trial.

It will not, I presume, be denied that the judge Advocate, from the constitution of this Court combines the two characters of judge and juror, and that it follows, as an indisputable consequence, that my objection which applies to either character is strictly applicable to him.

All therefore that remains for me to do, is to lay before you the legal authorities, on which I found my right of challenge.

There follow eight authorities that, while of little general interest, show that Macarthur possessed a small legal library. The speech concludes as follows.

Gentlemen, it would be an unpardonable waste of your time, and an insult to your understandings, to press upon you more authorities, for those I have submitted are clear to the point.

You will now decide, gentlemen, whether Law or Justice shall finally prevail over the contrivances of George Crosley, you have the eyes of an anxious public upon you, trembling for the safety of their property, their liberties, and their lives.

To you has fallen the lot of deciding a point which involves perhaps the happiness or misery of millions yet unborn, and I conjure you, in the name of the Almighty God, in whose presence you stand, to consider the inestimable value of the precious deposit with which you are entrusted.

For my own part, knowing you as I do, I have no apprehensions. I feel assured, that neither expectations of reward, and favour, nor dread of persecution, will influence your decision.

It is to the Officers of the New South Wales Corps that the administration of justice is committed, and who that is just has anything to dread.

JOHN MACARTHUR.

After Macarthur's protest was heard, Atkins demanded he be gaoled and Captain Anthony Fenn Kemp closed down the court. Bligh's efforts to make it reconvene, and the officers' response, led to the rebellion the next day. (For the indictment against Macarthur see HRNSW vol. 6 p. 467. For his speech see the same volume pp. 422ff.)

Acknowledgments

This book arose from my curiosity about two linked mysteries: the causes of the so-called Rum Rebellion and the character of John Macarthur. I had been told by teachers at school and university, and by a dozen popular writers, that Macarthur brought on the rebellion because Bligh interfered with his income from spirits. However, I later learned that historians had disproved this before I was born, which introduced a third mystery: Why had the rum myth persisted despite the evidence? I came to suspect it was because of some need to see Macarthur as bad. He was an extreme character, a man of great talents and enthusiasm for controversy, who achieved much in the areas of business and politics. I can think of few of my own contemporaries who come near him: possibly Malcolm Turnbull would be closest. Like Macarthur, he excites people, and he confuses them.

To learn more of Macarthur, I read the original documents relating to the rebellion in the *Historical Records of New South Wales* and was struck by something completely unexpected, the frequency with which those involved wrote and talked of honour and their own status as gentlemen, particularly during their disputes. I decided to write a book putting such words into their historical context, to see if this might provide an alternative explanation for the rebellion.

I relied on the work of many historians and economists to do this, and acknowledge the most important in the introduction to Sources.

I am grateful to George Parsons, John Wrigley and John Gascoigne for talking to me about Macarthur, to Bruce

Kercher and Kirsten McKenzie for some helpful comments, and to Alan Robinson, Nicolas Rothwell, Alex Snellgrove, Max Suich, and Anne-Maree Whitaker for reading the manuscript and telling me what they thought. Alan Robinson also smoothed access to Camden Park and John Macarthur Stanham and his wife Edwina kindly allowed me into their house. Edgar Penzig told me all I know about guns. Alex Snellgrove did the cover with Dolores Haynes, and Maggie Cooper was the typesetter, all magnificent as ever. Thanks to Ross Gibb, Peter Phillips, Tom Gilliatt and Katie Williams and her staff at Pan Macmillan. Carl Harrison Ford was a wonderful editor, wide and deep, and Neil Thomas improved the words and made many useful suggestions.

The staff at the Mitchell Library – including Helen Harrison in the pictures section – were most helpful. I particularly thank Paul Brunton, the library's senior curator, for a talk he gave some years ago on the loss of the *Guardian* after it was struck by an iceberg. For some reason I was fascinated by his story of how the ship's log book shows the decline in Lieutenant Riou's hand-writing after his right arm was broken by a falling barrel. Paul Brunton was kind enough to allow me to look at the log book, held in the library's manuscripts collection, and the sight of the alteration in the young commander's script touched me and rekindled my enthusiasm for early Australian history.

Picture & Map Credits

Pictures from the National Library of Australia, in order of appearance. Dates show creation or first publication where known, all call numbers begin with nla.pic-an.

Reproduced in black and white: George III 1791 by G.S. Facius 7200568; Arthur Phillip 1789 by W. Sherwin 9846227; Ingleses en la Nueva Holande 1793 by Felipe Bauza DGD2f482; John Hunter 1792 by Daniel Orme 9631633; Sir Joseph Banks 1801 by K. Mackenzie 928328; Sydney 1802 by W.S. Blake 6016088; William Bligh 1814 by Alexander Huey 11230917; Lachlan Macquarie 1895 14182923.

Reproduced in colour: Government House 1791 by William Bradley 3329075-1; The Residence of John Macarthur Esqre (Elizabeth Farm) c1822 by Joseph Lycett 7690900; four pictures by John Hunter 1788-1790, kangaroo 3172052, eastern rosella 3149498, emu 3186920, kookaburra 3148774; Convict Uprising at Castle Hill 1804 5577479; Arrest of Governor Bligh 1808 7655522.

Pictures from the Mitchell Library in the State Library of New South Wales.

Reproduced in black and white: Samuel Marsden GPO1-05189; Philip Gidley King and Family ML1244; William Paterson DG175; George Johnston GPO1-06755;

Reproduced in colour: John Macarthur GPO1-06190; Elizabeth Macarthur GPO2-06724; Government House Parramatta 1805 by G. Evans SSVIB/PARR/12; Sydney 1808 by John Eyre XVI/1808/9; Sydney 1808 by John Eyre XVI/1808/8; Government House Sydney 1807 by John Eyre SU/31.

Juan Ravenet's picture of the reception for the Spaniards (detail) is in the Museo de América, Madrid. The maps were drawn by Alexandra Snellgrove using a range of original sources and, for the map of Sydney in 1808, a modern map done by Bryan Thomas in 1979 (a copy is held in the Mitchell Library), for which the author is most grateful.

Notes

Epigraph

Bligh to Greville, 5 November 1807; Banks papers, 6, cited in Atkinson 1971
me than existence Marshall's trial HRA series I vol. 3 p. 216

Introduction

1 **their own initiative** The Reverend Richard Johnson and his wife preceded them, but only after he had been offered the job of chaplain to the First Fleet and urged to accept it by Wilberforce. And they did not stay.

5 **year in Britain** Kiernan pp. 143–144 quotes F. Wilkinson's *The Illustrated Book of Pistols*, London 1979, that of 200 British duels examined, one in six produced a wound, and one in fourteen a fatality. At p. 410 (note) Andrew quotes these estimates of the annual number of duels around this time: 33, 69, 2.6 and 2.9. Kiernan quotes one estimate of 172 duels during the long reign of George III (the same as the 2.9 figure given by Andrew) of which 91 had fatal results, 18 came to trial, and two killers were executed. Clearly, the variance between some of these figures is considerable.

8 **the other rebels** Some examples picked at random: Robert Hughes in *The Fatal Shore* (London 1987) writes 'Colonial Sydney was a drunken society, from top to bottom. Men and women drank with a desperate, addicted, quarrelsome single-mindedness' (p. 110); 'nearly all the men and most of the women were addicted to alcohol' (p. 290). He describes the New South Wales Corps simply as a 'junta' and has Macarthur organising the rebellion from his prison cell (p. 111). Johnston he dismisses as 'naively seditious' (p. 294). In *Sydney* (Sydney 1999) Geoffrey Moorhouse explains the rebellion by Macarthur's 'disputatious nature' and his 'greed: his own greed and that of virtually the whole officer class in New South Wales' (p. 89). Under Grose the colony underwent 'a form of moral collapse' (p. 89) and Macarthur was the 'driving force behind this rape of the colony' (p. 90).

8 **William Howitt** William Howitt, *Land, Labour and Gold; or, Two Years in Victoria*, 1855. This is the first use of the term *rum rebellion* identified in the *Australian National Dictionary*, Melbourne 1988. I would be interested to hear from any reader aware of an earlier use (*michael@duffyandsnellgrove.com.au*)

8 **in the rebellion** Diary of Mrs Macquarie, Mitchell Library, quoted in Mackaness p. 262

I Voyage

For this chapter I am indebted to Alan Atkinson's work on John Macarthur's early life in his 'John Macarthur Before Australia Knew Him', Michael Flynn's book *The Second Fleet*, and the journal and letters of Elizabeth Macarthur. Regrettably, this journal is the only private record of any length in existence from John or Elizabeth. Arthur Bryant's *The Years of Endurance* (London 1942) was drawn on for my description of England. While overly positive, it provides a useful contrast to the typical view of Georgian Britain found in many books about early Australia.

12 **13 August 1766** Atkinson 1979 for this and other details of Macarthur's early life. He corrects M.H. Ellis on a number of points and adds much new information. He revises Macarthur's date of birth, previously thought to be 1767. On p. 29 at note 46 he

writes: 'For the day of his birth, unpublished before this, see [James Macarthur] "Memorandum for life of the late John Macarthur, Esq." n.d.; for the year, see James Macarthur's memoir of his father (1853?), p. 179, which is confirmed by his obituary in the *Sydney Herald*, 17 April 1834, and by the inscription on his tomb at Camden Park. He was baptised on 3 September 1767, a delay easily explained by the laxness of a sea-port parish. If born in 1767 he would have been too young to be commissioned in 1782, the minimum age being 16.'

14 ***to the bar*** Onslow p. 1

15 **criminals at home** the early chapters of Hirst are important for the evangelicals

15 **the First Fleet** Hirst p. 16

16 **upper class Britons** ibid p. 16

16 ***upon the earth*** quoted in E.S. Turner's *What the Butler Saw*, London 2001, p. 39

16 **14 August 1766** Hazel King, *Elizabeth Macarthur and Her World*, Sydney 1980, p. 3

17 ***indolent and inactive*** Onslow p. 52

18 **John was not** see Daniel Pool's *What Jane Austen Ate and Charles Dickens Knew*, New York 1993, p. 44

20 ***from being used*** Morrison quoted in Mackaness p. 124

20 ***not paying attention*** ibid p. 126

20 ***through Endeavour Straits*** ibid p. 127

21 ***all the voyage*** ibid p. 139

22 ***the greatest tyrants*** quoted in Dening, pp. 123–4; Barrow was connected to one of the mutineers and therefore not unprejudiced, but Dening notes that other early commentators saw the mutiny in similar terms

23 ***of God's works*** HRNSW vol. 2 p. 745

23 ***planted, in it*** ibid p. 748

24 ***disappointment will follow*** Tench p. 75

25 ***have been seen*** HRNSW vol. 2 p. 626

25 ***other human society*** ibid p. 564

26 ***scruple or regret*** Onslow pp. 2–3

29 ***depended upon theatre*** Thompson 1974 p. 390

30 **with these ships** Statham p. 10

31 **of small boats** Humphreys p. 11

31 ***sending them here*** HRNSW vol. 1 pt. ii p. 176

33 **account of events** HRNSW vol. 2 pp. 427–431

36 ***duelling in Europe*** Smith p. 480

36 **duel was necessary** ibid pp. 122–123

36 **before a justice** David M. Walton, *A Legal History of Scotland*, Edinburgh 1996, p. 497

36 **punishable by banishment** ibid p. 501

37 **Parliament in 1641** Alan Harding, *A Social History of English Law*, Harmondsworth 1966, pp. 164–165

37 **courts put together** William Holdsworth, *A History of English Law*, London 1903, p. 580

37 **it through parliament** ibid p. 579

37 ***to undervalue him*** quoted in Andrew p. 411

37 ***him a coward*** quoted in the Addison chapter in *The Penguin Book of Fights, Feuds and Heartfelt Hatreds* (ed. Philip Kerr), London 1992

38 ***for military smartness*** quoted in George Parsons, 'The Commercialisation of Honour: early Australian capitalism 1788–1809' in Aplin

38 ***with the nimble*** Tobias Smollett, *Travels Through France and Italy*, 1766, quoted in Kerr op. cit. p. 58

39 **therefore less accurate** my thanks to Edgar Penzig for this fact.

39 **gentlemen in England** quoted in Andrew p. 414
39 ***company, in general*** quoted in Andrew p. 415
40 ***some distant hints*** HRNSW vol. 2 p. 488
41 ***in his ship*** ibid pp. 430–431
50 ***hundred per cent*** HRNSW vol. 1 pt. ii p. 366
51 ***at her anchors*** ibid p. 311
52 ***defray the cost*** ibid pp. 366–367
54 ***there was disease*** Sir James Watt, 'The Colony's Health', in Hardy and Frost

II Genesis

Alan Frost's two books, *Arthur Phillip, 1783–1814: his voyaging*, and *Botany Bay Mirages* are important reading for this period, as are several of the essays from the book edited by Frost and John Hardy, *Studies from Terra Australis to Australia*. The published journals of Watkin Tench and David Collins are full of incident. John Moore's *The First Fleet Marines 1786–1792* is helpful for details about the soldiery.

56 ***this delightful weather*** HRNSW vol. 1 pt. ii p. 368
60 ***in this fleet*** ibid p. 367
60 ***reached the shore*** ibid p. 387
63 **at Government House** Onslow p. 20
66 **Macassan trepang fishermen** Frost 1994, Ch. 10. Judy Campbell's *Invisible Invaders*, Melbourne 2002, a more recent consideration of the subject, supports this hypothesis.
67 ***and deserted victims*** D.D. Mann, *The Present Picture of NSW*, London 1811
67 **honour might proceed** Ralph Clark, *The Journal and Letters of Lieutenant Ralph Clark 1787–1792* (eds Paul Fidlon and R.J. Ryan), Sydney 1981, pp. 149–155
69 **Captain George Johnston** There is some confusion in Clark's diary over the identity of two men he names Johnstone. Clark was a poor speller. Captain Johnstone was in fact Captain-Lieutenant George Johnston.
73 ***what has passed*** Clark p. 280
76 **of the time** Sir James Watt's 'The Colony's Health' in Hardy and Frost, p. 146
81 ***circulation of air*** HRNSW vol. 1 pt. ii p. 349
82 ***the mother country*** ibid p. 370
83 ***his Captain ship*** John Cobley (ed.), *Sydney Cove 1791–1792*, Sydney 1965, p. 101
84 **honour was satisfied** Johnson to Stonard, Johnson Papers (St Pauls, Melbourne), cited in Ellis p. 37
84 **He was cashiered** Harris Papers (Mitchell Library), Harris to Hill, quoted in Ellis p. 29
85 **to the press** Whitaker 2000, pp. 34–35
87 **some relevant descriptions** HRNSW vol. 2 pp. 793–796
89 ***is served out*** Grose to Matthew Lewis, 22 October 1792, quoted in Whitaker 2000, p. 39
90 ***such a disgrace*** HRA series I vol. 6 pp. 239–240
91 **and government agriculture** HRNSW vol. 1 pt. ii p. 654
91 ***I know of*** ibid pp. 672–673
93 ***or savage coast*** Arthur Phillip, *The Voyage of Governor Phillip to Botany Bay*, London 1789, p. 122

III Elizabeth Farm

Brian H. Fletcher's *Landed Enterprise and Penal Society* explains the importance and significance of the officers' control of most of the capital in early Sydney. D.R. Hainsworth's *The Sydney*

Traders describes the extent of their trading activities. N.G. Butlin's monograph 'Yo, Ho, Ho and How Many Bottles of Rum?' defends them against the charge of abusing the trade in spirits. His monograph 'Free Lunches Antipodean Style' sets out the degree to which the officers benefited financially from their unique situation. I have drawn on these authors' work so extensively for this chapter that I have not provided specific reference notes. George Parsons' essay 'The Commercialisation of Honour', in Aplin, is instructive on the strange culture of the New South Wales Corps.

102 **are of interest** see King
103 ***the ape family*** quoted in King p. 159
106 ***bartering spiritous liquors*** HRA series I vol. 6 p. 434
107 **evidence of this** Hainsworth pp. 31–32
111 ***thirty years past*** M. Boulton to Lord Hawkesbury, 17 April 1790, quoted in Jenny Uglow's *The Lunar Men*, London 2002, p. 66
111 **authority and ministers** Castles pp. 2–3
112 ***an awkward bashfulness*** quoted in Mason pp. 64–66
112 ***high and elegant*** W.M. Thackeray, *The Four Georges*, quoted in Mason
113 ***make a gentleman*** Edmund Burke's letter to William Smith, 29 January 1795, quoted in *Oxford Library of Words and Phrases*, vol. 1, Oxford 1981
115 **had in Europe** Patrick O'Farrell, 'Australian Religion' in Goldberg & Smith
116 ***been more civil*** HRNSW vol. 2 p. 273
117 ***as a gentleman*** ibid p. 299
118 **rural a scene** ibid p. 817
122 ***generation of addicts*** Sir James Watt, 'The Colony's Health', in Hardy & Frost, p. 151
123 **the other officers** In making this judgement I am accepting the view formed by John Hunter some time after he became governor. See for instance his despatch to the Duke of Portland, 25 July 1798, HRNSW vol. 3 pp. 418–430. It is a confusing business and we don't have much direct evidence for or against Macarthur, who denied those of Hunter's accusations he heard about. But Hunter had no reputation for lying, nor any reason to lie in this case, whereas we know that Macarthur fudged the truth on occasion. Macarthur's growing wealth, the apparent absence of other major sources of income, and the number and type of complaints made against him over the years, provide circumstantial evidence that he was involved in the sort of behaviour alleged by Hunter.
124 ***of their being*** Alexis de Tocqueville, *Democracy in America*, 1835, New York 1945, vol. 1 pp. 26–7
124 **criminals in abundance** for this and other observations on convicts I am indebted to John Hirst's *Convict Society and its Enemies*
125 **in Australia today** Frost 1987
127 **than in Britain** Gascoigne pp. 6–7

IV Hunter

This chapter is based closely on the documents in volume 3 of the *Historical Records of New South Wales*. As they can be easily identified, I have not provided details in most cases.

135 ***necessaries of life*** quoted in Bernard Smith (ed.), *Documents on Art and Taste in Australia 1770–1914*, Melbourne 1975, pp. 11–13
137 ***the other settlements*** Collins vol. 1 p. 375
143 ***come to market*** HRNSW vol. 3 p. 31
144 ***to this settlement*** ibid pp. 122–123

145 **it a go** ibid pp. 125–126
148 **increase of livestock** HRNSW vol. 6 pp. 131–136
151 **they were convicts** HRA series I vol. 1 pp. 587–589
151 **held by Foveaux** Some authors have said Macarthur became paymaster earlier than this, but see Whitaker 2000, p. 47
153 ***and his buffoonery*** Thompson 1974 p. 399
153 **offers of rewards** ibid
160 ***of his character*** Onslow pp. 46–52
160 **and invested it** Hainsworth p. 25
160 **face court martial** HRNSW vol. 3 pp. 636–638
161 ***proved considerably greater*** ibid pp. 673–675
163–4 ***an amusing novel*** quoted in AGL Shaw, 'Some aspects of the history of New South Wales 1788–1810', JRAHS vol. 57, June 1971
164 **him with flogging** T. Crofton Croker, (ed.), *Memoirs of Joseph Holt*, London 1838
165 **defend his interests** Evatt pp. 39–45
168 **certain emancipist traders** Hainsworth pp. 42–43
169 **the government herd** HRA I, 2, p. 505
169 **of the colony** HRA, I, 2, p. 672

V King

174 **any written instructions** HRNSW vol. 4 p. xxviii
182 **Town in 1797** Ellis p. 168
182 ***the coarser kind*** HRA I, 3, p. 563
182 **New South Wales** Atkinson 1997 p. 270
185 ***knock you down*** HRA series I vol. 3 pp. 199–216
186 **serve with him** Kiernan p. 101
187 ***claim, to honour*** HRA series I vol. 3 p. 215, quoted in Evatt p. 54
188 **fight the duel** for following account see HRA, Series I vol. 3 pp. 186–216; for Macarthur's belief that honour was relevant see his speech to the court p. 212
191 **guilty of assault** A strange aftermath of this was that on 25 September 1801 Macarthur apparently invited Jefferies to his house for dinner when Abbott was present. What significance – if any – this had is unknown. See HRA series I vol. 3 pp. xviii, 320
191 ***the royal mercy*** HRNSW vol. 4 p. 525
196 ***Whenever he pleases*** HRA series I vol. 3 p. 297
197 **ended without bloodshed** Kiernan p. 109
197 **from the army** Old Officer pp. 164–165
197 ***army with disgrace!*** ibid p. 166
198 ***out twelve steps*** The record says twelve feet but this would have been too close, so presumably twelve steps was meant
199 ***him to relax*** HRNSW vol. 7 p. 242
200 **by the traders** quoted in Peter Watson's *A Terrible Beauty*, London 2000, p. 171
201 ***with or for*** quoted in Eldershaw p. 39
202 ***to be amusing*** Ritchie p. 41
202 **of the duel** HRA series I vol. 3 p. 475
203 ***of valuable ones*** HRNSW vol. 4 pp. 564–565
203 ***oft times mend*** Ritchie pp. 42–44
204 ***the blackest treachery*** HRA series I vol. 3 p. 308
207 ***now worth £20,000*** HRNSW vol. 4 pp. 611–615
207 **evidence of this** Ellis pp. 206–207

208 **he been killed** Kiernan p. 8
211 ***here and elsewhere*** Piper Papers, A256, pp. 464–465, 4 December 1801, quoted in Ellis

VI Luck

213 **and some Europeans** Frank Horner, *Looking for La Perouse*, Melbourne 1995, p. 99
214 ***it.*** **And resign** Atkinson 1980 p. 78
214 **New South Wales** this account is based on the version in Atkinson 1980
214 **Stevensons and Hallidays** Atkinson 1971 p.29
214 ***his venal associates*** Piper Papers, A254, 8 September 1802, cited in Ellis
216 ***a military tribunal*** HRNSW vol. 5 pp. 11–13
216 ***too long already*** Piper Papers, A256, p. 443, cited in Ellis
216 **in the Pacific** Colin Forster, *France and Botany Bay*, Melbourne 1996, pp. 7–8
217 ***the dazzling vine*** Horner p. 9
217 ***of eight years*** Frost 1989 p. 224
218 **malnutrition and disease** Robert Darnton, *The Great Cat Massacre*, New York 1999, p. 27
218 **spirits were dropped** HRNSW vol. 4 pp. 393–394
219 ***their military duty*** ibid p. 992
219 **health was improved** HRA series I vol. 4 pp. 218–219
219 ***require absolute retirement*** HRNSW vol. 5 p. 140
220 **very small stock** ibid p. 63; see previous pages for other two letters
221 ***influence of government*** ibid p. 460
222 ***of useful discoveries*** Gascoigne p. 4
223 **Saxony and Holland** Garran & White
223 **very little expense** Carter: Fawkner to Banks p. 403
224 **at Botany Bay** Carter p. 404
224 ***our common enemy***: Macarthur to J. Piper, 9 November 1803, Piper Correspondence, 3, quoted in Atkinson 1971, p. 31
225 ***of the deceased*** Brian Fletcher, ed., *The Present Picture of New South Wales 1811*, by D.D. Mann, 1811, Sydney 1979
226 ***a large family*** quoted in Ellis p. 236
226 ***croppies had risen*** quoted in Eldershaw pp. 59–60
227 ***given me Greek*** quoted in Eldershaw p. 55
227 **all he wanted** Johnston to Piper, quoted in Eldershaw p. 56
229 **hanged or flogged** This event has traditionally been known as the Vinegar Hill or Castle Hill Rebellion or Uprising. But in an article by John Huxley in the *Sydney Morning Herald* of 22–23 February 2003, p. 5, it is referred to as the Castle Hill massacre.
229 ***vexation or other*** quoted in Eldershaw p. 54
229 **party of settlers** HRA series I vol. 5 p. 503
231 ***live in England*** HRNSW vol. 5 pp. 387–391
231 ***rich luxuriant grass*** Onslow p. 87
231 **rather than freehold** HRNSW vol. 5 p. 365
231 **the government changed** Atkinson 1971 p. 32
231 **produce fine wool** HRNSW vol. 5 p. 450
232 **convicts as shepherds** ibid pp. 370–373
232 **it was one-quarter** ibid p. 450
232 ***worth attending to*** ibid p. 396
232 **from then on** Evidence of this, apart from Hunter's words to the committee, was provided years later by Macarthur's son James, who recalled that his father and he often

visited Hunter on friendly terms in the period 1810–14.

232 *of sheep only* HRNSW vol. 5 p. 399

233 **colony, the Cowpastures** For discussion of Evatt's claim that Macarthur was not offered 10,000 acres see Atkinson 1971 pp. 37–38

233 **of several noblemen** Martin Mason letter to Sir Joseph Banks, copy in Bligh Papers, Mitchell Library, S1/45 p. 156, quoted in Paul Brunton's 'Arresting Bligh' in *Mutiny on the Bounty*, Sydney 1991

233 **beasts on sale** Onslow pp. 98–99

234 *debauched old rascal* Martin Mason letter op cit

237 **some 3,500 acres** This is the figure given in the governor's return of August 1805. Macarthur told the British government in July 1804 that he had about 4,000 acres.

237 **burst into tears** Hazel King, *Elizabeth Macarthur and Her World*, Sydney 1980, p. 37

238 **the whole country** HRNSW vol. 5 p. 669

238 *[be] ... found* ibid p. 49

238 *all becoming humility* ibid p. 332

240 **Parramatta Loyal Association** *Sydney Gazette,* 13 July 1805

242 **to come in** HRNSW vol. 6 p. 1

242 **conversation with King** C.H. Currey noted the link with Bentham in his *The Brothers Bent*, Sydney 1968, p. 10. Alan Atkinson (1978) developed the connection and suggested that Bentham could therefore be held responsible for the events of January 1808. Yet none of the rebels, including Johnston when he was on trial for his life, mentioned Bentham's opinion in connection with the rebellion.

243 *the rising generation* HRNSW vol. 6 p. 93

245 *his youth employed?* Carter p. 542

245 **Captain Henry Waterhouse** ibid p. 456

246 *kind of indulgence* ibid pp. 543–544

VII Bligh

248 **of the navy** Atkinson 1971 pp. 92–95

248 **of Bligh's appointment** Banks to Bligh, 21 April 1805; Banks papers, 21

249 **New South Wales** Whitaker 2000 pp. 86–87

251 **sold other property** O'Brien p. 231

252 **an enormous mark-up** Ritchie pp. 83–84

252 **fashion towards him** HRA series I vol. 6 pp. 554–555

253 *of the Crown* HRNSW vol. 6 p. 572

253 **shillings a gallon** Hainsworth pp. 56–57

253 **to extract confessions** Ritchie p. 98

254 *her parent's head* E. Bent Papers, FM3/20, p. 162, cited in Ellis

256 *they had witnessed* Ritchie pp. 178–180. Against the veracity of this account is that Macarthur remained on polite terms with Bligh for some time afterwards. In its favour is that Bligh opposed large land grants and knew Macarthur had lied to get his.

256 *of his reign* Onslow p. 137

257 *his launch again* Ritchie p. 136

257 **from the Cowpastures** HRNSW vol. 6 p. 248

257 *never seen again* ADB vol. 1 p. 573

258 **a sealing vessel** Hainsworth p. 159

258 **proposing to form** HRA series I vol. I p. 160

259 **attacks on Bligh** Atkinson 1971 pp. 100–101

261 **them to transportation** B.W. O'Dwyer, 'Michael O'Dwyer and the 1807 Plan of Insurrection', JRAHS vol. 69 pt. 2 p. 77

261 **system was concerned** see also letter from John Harris to Philip Gidley King of October 1807, HRNSW vol. 5 p. 338

261 **despatch to London** HRA series I vol. 6 p. 159

261 **fashionable Picturesque manner** Tim McCormick (ed.), *First Views of Australia 1788–1825*, Sydney 1987, p. 119

261 ***is your habitation!*** Ritchie p. 365

262 **to Major Johnston** 'Sergeant-Major Whittle's Statement', HRNSW vol. 6, p. 588. At Johnston's trial, Bligh denied ordering Whittle to do this (Ritchie p. 394). As Johnston included Whittle's claims in despatches to London, it would have been an unusually elaborate lie.

262 **maybe ten people** Alan Atkinson, 'Taking Possession: Sydney's first householders', in Aplin

262 ***other against him*** Ritchie p. 352

262 **would be next** Atkinson 1997 p. 280

262 **person of £2,000** Whitaker 2000 p. 109

264 ***such a government*** Ritchie p. 217

264 ***an unpleasant sensation*** Ritchie p. 351

265 ***by those laws*** quoted in Humphreys pp. 99–100

265 ***and possessing property*** Andro Linklater, 'Life, Liberty and Property', from the magazine *Prospect* August 2002, pp. 28–30, based on his book *Measuring America*, New York 2002

265 ***of a governor*** Ritchie pp. 386–87 cited in Atkinson 1971

267 **and other epithets** Ritchie p. 375

267 ***of the moment*** Ellis p. 283 note 24

267 ***consider themselves gentlemen*** Bligh to Greville, 5 November 1807; Banks papers, 6, cited in Atkinson 1971

267 **the free residents** Ellis p. 284

267 ***the military officers*** Bligh Correspondence, Mitchell Library, safe 19, No. 7: Mrs Putland to Elizabeth Bligh, 10 October 1807, cited in Ellis

268 ***will be well*** quoted in Mackaness vol. 2 p. 126

269 ***New South Wales*** HRNSW vol. 6 p. 335

269 **huge symbolic importance** Macarthur strenuously denied any suggestion he had used the courts for political purposes, no doubt aware that this would look bad back in Britain where people were unaware of colonial ways and needs. See his cross-examination of Edmund Griffin HRNSW vol. 6 pp. 482–485 regarding his comments before the magistrate's court in October 1807.

269 ***no independent press*** Evatt p. xv

270 **had been *inflammatory*** HRNSW vol. 6 p. 484

270–1 ***not stand trial*** ibid pp. 491–492

271 ***as Mount Ararat*** Ritchie p. 217

271 ***be an alteration*** Ritchie p. 136

271 **and his daughter** Ellis p. 290

272 ***much wounded also*** *Sydney Gazette* 16 November 1806

272 **in October 1807** HRA series I vol. 6 pp. 159–160

274 **honour could tolerate** HRNSW vol. 6 p. 509

275 ***objects of prey*** Thompson 1978 p. 139

277 ***openly spoke of*** HRNSW vol. 6 p. 553

277 **them in court** Ritchie p. 67

277 ***to private inclination*** ibid p. 60

277 **accessible to everyone** ibid p. 37

277 ***town of Sydney*** ibid p. 188

278 **in mid-January 1808** HRNSW vol. 6 pp. 831–832
279 **be a Freemason** *Sydney Gazette* 25 May 1803
279 **the generous Macarthur** Ritchie p. 371
280 **the day's significance** I'm grateful to Anne-Maree Whitaker for this observation
280 ***admirer* of Fox** Onslow p. 471

VIII Rebellion

The main versions of the rebellion are in George Johnston's despatches, various accounts by Bligh, and the evidence at Johnston's trial transcribed in *A Charge of Mutiny*. Often the accounts differ. The most comprehensive marshallings of the sources are in George Mackaness' *The Life of Vice-Admiral Bligh* and in Ross Fitzgerald and Mark Hearn's *Bligh, Macarthur and the Rum Rebellion.*

281 **who at sixty-three** HRNSW vol. 6 p. 490 footnote
282 ***not to sit*** Ritchie p. 219
282 **attack on Atkins** HRNSW vol. 6 pp. 422–425
283 **drinking with Crossley** Onslow p. 147 note
283 **died in 1797** My thanks to Anne-Maree Whitaker for this observation
284 ***for political purposes*** cited in Atkinson 1980 note 78
284 ***you*, Kemp responded** Ellis p. 332
284 ***ruffians – the police*** HRA series I vol. 6 p. 239
285 ***their lives dearly*** Onslow p. 151
285 ***surgeon Thomas Arndell*** Ritchie p. 166
286 **or organised religion** quoted in Gascoigne p. 22
286 **presented to Bligh** Ritchie p. 167
286 ***ordered to gaol*** Ritchie p. 220
286 ***he arrest us*** Statement of John Colonan, a soldier, [1811?] War Office 72/35, quoted in Atkinson 1997
287 **had him transported** O'Dwyer p. 78
287 **them in gaol** HRA I 3 pp. 231–233
287 **the next day** HRNSW vol. 6 pp. 489–490
288 **for the riots** Thompson 1974 p. 397
289 **the Machiavellian Macarthur** Captain Edward Abbott claimed Johnston had 'suffered himself … to be led' but he was referring to the aftermath of the rebellion, not its occurrence. He supported it and believed that had Bligh not been arrested 'there would have been a mutiny; there is no doubt of it', not least because Bligh had supported Gore in appointing a large number of Irish constables who insulted the soldiers in Sydney. See HRNSW vol. 6 p. 834.
290 ***as a magistrate*** Ritchie p. 155
291 **the male population** Statham p. 11
291 **houses or cattle** Ritchie pp. 279 and 352
291 ***certainly take place*** ibid p. 152
292 ***made away with*** ibid
293 **the public consciousness** Whitaker 2000 p. 23
293 **from a prisoner** ibid 2000 p. 89
293–4 **room with Johnston** Mackaness vol. 2 p. 185
294 **my own motives** Ritchie p. 212
295 ***into his house*** Boswell's *Life of Johnson* (1783), vol. 3 p. 245
296 ***pretend to say*** Ritchie pp. 241–242
296 ***the British Grenadier*** My thanks to Anne-Maree Whitaker for these words

296 ***will be killed*** HRA series I vol. 6 p. 549
296 **reviewed in Whitehall** Ellis p. 350
298 **ingenious manner described** Ritchie pp. 371–373
298 ***down stairs together*** ibid p. 243
299 ***on his coat*** ibid p. 329
299 **to be murdered** HRNSW vol. 6 p. 617
301 **as they unfolded** Alan Atkinson, while acknowledging that the rebellion was 'partly designed to avenge his [Macarthur's] fragile dignity', has described it as a 'calculated' affair on Macarthur's part, indeed 'a safe one'. He has suggested Macarthur would not have acted unless he believed his friends in Britain would protect him (Atkinson 1971 p. 28). Yet the composition of the government was unpredictable.
301 ***I ever saw*** Banks papers, 6, quoted in Atkinson 1971 p. 29

IX Aftermath

Copies of the letters from Macarthur to his wife are found in chapter six of *Some Early Records of the Macarthurs of Camden* edited by Sibella Macarthur Onslow while the trial transcript is in *A Charge of Mutiny* introduced by John Ritchie.

303 ***his own emolument*** HRNSW vol. 6 p. 461
304 ***breed of sheep*** *Morning Chronicle*, 14 November 1807, quoted in Whitaker 2000 p. 98
304 ***to defend him*** Elizabeth Bligh to Banks, 14 January 1808, HRNSW vol. 6 pp. 417–418
305 ***to Norfolk Island*** Mackaness pp. 106–110
305 ***enraged against me*** Ritchie p. 93
306 **for several weeks** ibid p. 120
309 **celebrate his acquittal** ibid p. 303
309 **with great energy** HRNSW vol. 6 p. 513
309 **defeat James II?** Humphreys p. 56
309 **on 26 January** Ritchie pp. 300–301
310 ***of his enemies*** HRNSW vol. 6 p. 513
310 **and the rebellion** Ritchie p. 300
310 ***about her recovery*** Ellis p. 571 note 28 for source and discussion of dating of letter
310 **the government's bills** Ellis p. 369
311 **other leading rebels** Atkinson 1978. He points out that the document, incorporating Bentham's belief in the illegality of much of the governor's power, 'did find its way to the Colonial Office', but it seems it was never produced publicly. He suggests this 'chief justification for the rebellion' was not used during Johnston's trial because its concerns were irrelevant to the court martial. But this would not explain why, if it was so potent, it was not used earlier by the rebels in their efforts to gain political and public support. Atkinson suggests this was because it implicity contradicted their justification for the rebellion, which was that Bligh had acted illegally. My own feeling – and the evidence is so slender it's impossible to be more precise than this – is that Bligh's illegality was a relatively small part of the total of justifications made by the rebels and accepted by the court.
312 **to overthrow Bligh** Hainsworth p. 70
312 **India Company's monopoly** ibid p. 172
312 **back to Sydney** HRNSW vol. 6 p. 678
314 ***deponent cannot tell*** HRA series I vol. 6 p. 397
314 ***cut you down*** HRNSW vol. 6 p. 691
315 **be his superior** Whitaker p. 110
315 ***to the government*** quoted in A.T. Yarwood's *Samuel Marsden: the great survivor,*

Melbourne 1977, p. 117
315 ***always will encounter*** Onslow pp. 164–165
316 ***foe to democracy*** Tench, *Letters from Revolutionary France* (ed. Gavin Edwards), Cardiff 2001, p. 100
317 **a powerful Whig** Atkinson 1971 p. 105
317 **sympathised with Bligh** ibid p. 106
318 **colonial secretary's position** Whitaker 1998 p. 73
318 ***you may determine*** Macarthur to Johnston 16 January 1809, Dixson Library MS Q22 Johnston papers
319 **Wales gentry were** *Sydney Gazette,* 15 January 1809 to 2 April 1809
320 **Bligh and Paterson** Ritchie p. 216
320 ***must be drawn*** Ritchie p. 158
321 **government and public** Parsons in Aplin p.115
322 ***he enters into*** HRA series I vol 7 p. 331
323 **failed trading ventures** Hainsworth records on pp. 174–175 that Macarthur even brought 110 tons of sandalwood on the voyage from Sydney to London for Johnston's trial, disposing of it to an American captain in Rio for a better price than was being offered in Canton. Even so, according to Hainsworth, he probably lost money overall on his sandalwood ventures.
326 **in November 1811** Although there has been no suggestion that the transcript is inaccurate or incomplete, on average there are only 37 pages per day, including documents provided to the court.
326 **by top lawyers** Evatt p. 342
326 **in this barter** Richie p. 19
326 ***to the 26th*** ibid p. 82
327 **an infant colony** ibid p.140
327 ***revolution is unavoidable*** Onslow p. 195
327 **of arresting them** Ritchie pp. 400–401
327 **and most alarming** ibid p. 383
328 ***in my life*** ibid p. 60
328 **Macarthur to move** ibid p. 393
328 **from their houses** ibid pp. 78 (Palmer), 98 (Gore)
328 ***and disgraceful cowardice*** Onslow p. 210
328–9 ***boil with indignation*** Ritchie p. 153
329 ***others to determine*** ibid p. 391
329 ***war or battle*** Mervyn James, *English Politics and the Concept of Honour 1485–1642*, Past & Present Supplement 3, Oxford 1978, p. 91
329 ***moralisation of politics*** ibid p. 2
330 **was an invention** Lang vol. 1 p. 124
330 **drunk on duty** Whitaker 2000 p. 146
331 ***forfeit a life*** Diary of Mrs Macquarie, Mitchell Library, cited in Mackaness p. 262
331 ***of Governor Bligh*** the *Times*, 5 July 1811, p. 2e, quoted in Whitaker 2000 p. 146
331 **returned to Sydney** see recommendation of T.G. Harris, legal advisor to the Crown, HRNSW vol. 7 p. 213
332 **such as *farce*** Ritchie, p. viii
332 ***onto something*** Robert Darnton, *The Great Cat Massacre*, New York 1999, p. 5
333 ***and obey them*** quoted in Humphreys p. 194

Epilogue

336 ***calculation at defiance*** Onslow pp. 215–216

Sources

The original part of this book is based on the words of Macarthur and those he dealt with, most of which are contained in the *Historical Records of New South Wales*, the *Historical Records of Australia*, and *A Charge of Mutiny*, the transcript of George Johnston's trial. These collections, together with Sibella Macarthur Onslow's *Some Early Records of the Macarthurs of Camden*, also contain the majority of the known significant facts of Macarthur's life. For other facts, and to understand the context in which his words and life need to be seen, I am grateful to the authors of many books, particularly two by inspired amateur historians. H.V. Evatt's *Rum Rebellion* and M.H. Ellis' *John Macarthur* are so extraordinarily biased – one against him and one for him – that the reader can allow for this easily and enjoy the brilliance that their writers bring to their subject. Ellis' book, the only previous biography, is a masterpiece of original research and the general shape of Macarthur's life it establishes is still largely accurate, despite all the history that has been written since 1955.

As indicated in the Acknowledgments, I am indebted to two modern historians. George Parsons' essay 'The Commercialisation of Honour: early Australian capitalism 1788–1809' (in *A Difficult Infant: Sydney before Macquarie*, edited by Graeme Aplin) is the only work I'm aware of to suggest that honour played a part in the Bligh rebellion. Apart from Parsons' own ideas I found some of his sources very helpful to my understanding of the code of honour. And the most important knowledge about Macarthur added since Ellis is from Alan Atkinson's research in British archives. In particular he has fleshed out our knowledge of Macarthur's life just before he came to Australia, and the pattern of patronage that played such an important part in his subsequent success. Most of this work remains in an unpublished MA thesis and in articles in academic journals, and I am fortunate to have been able to draw on it.

What follows is a select bibliography, of those books and articles that have contributed significantly to my understanding of Macarthur's life. My favourites, apart from the above, include George Mackanness' biography of Bligh, which contains the best account of the 1808 rebellion, John Hirst's *Convict Society and its Enemies*, Alan Frost's books, and Anne-Maree Whitaker's biography of Joseph Foveaux with its thorough treatment of patronage. On the gentleman and the code of honour I recommend V.G. Kiernan's book *The Duel in European History*, and Donna T. Andrew's article 'The Code of Honour and Its Critics: the opposition to duelling in England, 1700-1850'.

Sources of less relevance, and which have been quoted fewer than three times, are to be found in the endnotes. However, this is not an academic book so, in order to save space, where a source is obvious from the main text I have not provided an endnote. In particular, I have assumed that readers wanting to look up the correspondence between governors and London will be able to do this easily enough, in either of the historical records series.

Selected Sources

Abbreviations

Australian Dictionary of Biography	ADB
Historical Records of Australia	HRA
Historical Records of New South Wales	HRNSW
Journal of the Royal Australian Historical Society	JRAHS

The Macarthur and other papers in the Mitchell Library, Sydney

Books

Abbott G.J. and Nairn N.B. (eds), *Economic Growth of Australia 1788–1821*, Melbourne University Press, Melbourne 1969

Aplin, Graeme (ed.), *A Difficult Infant: Sydney before Macquarie*, Kensington 1988
Atkinson, Alan, *The Europeans in Australia: a history, Volume I*, Oxford 1997
Britton A. and Bladen F.M. (eds), *Historical Records of New South Wales* vols. I–VII, Government Printer, Sydney 1892-1901
Brunton, Paul, 'Arresting Bligh', in *Mutiny on the Bounty*, State Library of NSW, Sydney 1991
Butlin, S.J., *Foundations of the Australian Monetary System*, Melbourne 1953
Castles, Alex C., *An Introduction to Australian Legal History*, Sydney 1971
Collins, David, *An Account of the English Colony in New South Wales*, [1798, 1802] ed. Brian Fletcher, two volumes, Sydney 1975
Dening, Greg, *Mr Bligh's Bad Language*, Cambridge 1992
Dunmore Lang, John, *An Historical and Statistical Account of New South Wales Both as a Penal Settlement and as a British Colony*, 2 vols, London 1834
Eldershaw, M.B., *The Life and Times of Captain John Piper*, Sydney 1939
Ellis, M.H., *John Macarthur*, Sydney 1955
Evatt, H.V., *Rum Rebellion*, Sydney 1938
Fitzgerald, Ross and Hearn, Mark, *Bligh, Macarthur and the Rum Rebellion*, Sydney 1988
Fletcher, Brian H., *Landed Enterprise and Penal Society*, Sydney 1976
Flynn, Michael, *The Second Fleet: Britain's grim convict armada of 1790*, Sydney 1993
Frost, Alan, *Arthur Phillip, 1783–1814: his voyaging*, Melbourne 1987
Frost, Alan, *Botany Bay Mirages*, Melbourne 1994
Frost, Alan, *Convicts and Empire*, Melbourne 1980
Garran, J.C. and White, L., *Merinos, Myths and Macarthurs*, Canberra 1985
Gascoigne, John, *The Enlightenment and the Origins of European Australia*, Cambridge 2002
Goldberg S.L. and Smith, F.B. (eds), *Australian Cultural History*, Melbourne 1988
Hainsworth, D.R., *The Sydney Traders*, Melbourne 1981
Hardy, John and Frost, Alan (eds), *Studies from Terra Australis to Australia*, Australian Academy of the Humanities, Canberra 1989
Hirst, J.B., *Convict Society and its Enemies*, Sydney 1983
Humphreys, A.R., *The Augustan World*, London 1954
Kiernan, V.G., *The Duel in European History*, Oxford 1988
King, Robert J. (ed.), *The Secret History of the Convict Colony: Alexandro Malaspina's report of the British settlement of New South Wales*, Sydney 1990
Mackaness, George, *The Life of Vice-Admiral Bligh*, 2 vols, Sydney 1931
Macarthur Onslow, Sibella (ed.), *Some Early Records of the Macarthurs of Camden*, Sydney 1914
Mason, Philip, *The English Gentleman*, London 1982
Moore, John, *The First Fleet Marines 1786-1792*, Brisbane 1987
Old Officer, An, *Cautions and Advices to Officers of the Army, particularly subalterns, to which are added a letter, &c. on duelling*, Perth (Scotland) 1795

Pike, Douglas (general ed.), *Australian Dictionary of Biography*, vols 1 & 2, Melbourne 1967
Ritchie, John (intro.), *A Charge of Mutiny*, Canberra 1988
Statham, Pamela (ed.), *A Colonial Regiment*, Canberra 1992
Tench, Watkin, *A Narrative of the Expedition to Botany Bay* and *A Complete Account of the Settlement at Port Jackson*, in *Sydney's First Four Years* (ed. L.F. Fitzhardinge), Sydney 1961
Whitaker, Anne-Maree, *Distracted Settlement: New South Wales after Bligh from the Journal of Lieutenant James Finucane 1808–1810*, Melbourne 1998
Whitaker, Anne-Maree, *Joseph Foveaux*, Sydney 2000

Articles
Andrew, Donna T., 'The Code of Honour and Its Critics: the opposition to duelling in England, 1700-1850', in *Social History*, vol. 5 no. 3
Atkinson, Alan, 'The British Whigs and the Rum Rebellion', *JRAHS*, vol. 66 pt. 2, September 1980
Atkinson, Alan, 'Jeremy Bentham and the Rum Rebellion', *JRAHS*, vol. 64 pt.1, June 1978
Atkinson, Alan, 'John Macarthur Before Australia Knew Him', *Journal of Australian Studies*, no. 4, Melbourne, June 1979
Atkinson, Alan, 'A New John Macarthur', *The Push from the Bush*, no. 17, April 1984
Auchmuty, James Johnston, 'The Background to the Early Australian Governors', *Historical Studies Australia and New Zealand*, vol. 6 no. 23, November 1954
Butlin, N.G., 'Free Lunches Antipodean Style' No. 57 in *Working Papers in Economic History*, Canberra 1985
Butlin N.G., 'Yo, Ho, Ho and How Many Bottles of Rum?', *Australian Economic History Review*, vol. XXIII no. 1, March 1983
Campbell, Walter S., 'The Use and Abuse of Stimulants in the early Days of Settlement in New South Wales', *JRAHS*, vol. 18 pt. 2, 1932
Gallagher, J.E., 'The Revolutionary Irish 1800-1804, *The Push from the Bush*, no. 17, April 1984
Kemp, Murray C. and Kemp, Therese B., 'Captain Anthony Kemp', *JRAHS*, vol. 51 pt. 1, March 1965
Shaw, A.G.L., 'Some Aspects of the History of New South Wales 1788–1810', *JRAHS*, vol. 57 pt. 2, June 1971
Thompson E.P., 'Eighteenth Century English Society: class struggle without class?', *Social History*, May 1978, pp. 133–165
Thompson E.P., 'Patrician Society, Plebian Culture', *Journal of Social History*, Summer 1974, pp. 382–405

Unpublished Thesis
Atkinson, Alan, *The Position of John Macarthur and His Family in New South Wales before 1842*, MA thesis, University of Sydney, 1971

Index

Abbreviations M: Macarthur; BR: Bligh rebellion; JT: trial of Johnston.